NINETY-DAY WONDER

NINETY-DAY WONDER

HOW THE NAVY WOULD HAVE BEEN BETTER OFF WITHOUT ME

STEPHEN DAVENPORT

H.H. BONNELL

ISBN 9780976925545 (paperback) | 9780976925552 (ebook) | 9780976925569(audio)

Cover Images: USS *Vermilion* (AKA-107). Underway at sea, circa the 1950s. Official U.S. Navy Photograph, from the collections of the Naval History and Heritage Command. *Clouds*: Wahyu Ramdani/Shutterstock.com

First edition 2023

Book and Cover Design by Rachel Lopez Metzger

H.H. Bonnell, Publisher

To Joanna.

CHAPTER ONE

Every time I pass the exit from the Bay Bridge to Treasure Island on my way to San Francisco from Oakland, where I live now, I hear Kate Smith, otherwise known as the First Lady of Radio, singing. Her voice came out of the radio of a taxicab delivering me to Reserve Officer Candidate School at the Navy base that was on the island then, in 1951, the summer after my sophomore year at Oberlin. I was about to be incarcerated for eight weeks of training to convert me into an officer and gentleman, an absurd proposition, and I was sick with lovestruck loneliness for my girlfriend, Joanna Thompson, a freshman at Oberlin. She was on an island too that summer so long ago, Nantucket, another word for summer paradise, waiting tables by day, partying all night under bright stars on beaches to the sound of surf. I could only pray she wouldn't lose her heart to some ultra-smooth draft dodger with a shiny new convertible and a trust account who went to Yale. Now, in 2023, Joanna is ninety and I am ninety-two. We've been married since 1954.

"Hello young lovers, whoever you are,
I hope your troubles are few.
All my good wishes go with you tonight,
I've been in love like you.

Be brave, young lovers, and follow your star,
Be brave and faithful and true.
Cling very close to each other tonight,
I've been in love like you."

Nineteen fifty-one was the second year of the Korean War. In those long-ago days, before athletes specialized in only one sport, I played football in the fall and, until my junior year, basketball in the winter, and ran track in the spring. The coaches, most of whom had fought in World War II, several in the Navy, were afraid my teammates and I would be drafted away from their teams. To ensure we would stay in college for all four years, they suggested we sign up for the Naval Reserve Officer Candidate School, which required two summers of training, one after sophomore and one after junior year. If we passed, then we would be automatically commissioned as ensigns, the lowest officer rank, upon graduating from college.

We had no reason to distrust our coaches' advice. With very few exceptions, they were much closer to us than our professors. Twelve of us signed up—but not our friend and teammate Joe Howell. It didn't occur to us that the reason he didn't apply to the program was that he was sure he'd be turned down because of his race, African American—or Negro, in the parlance of the time. We just assumed he didn't want to sign up. It is hard to believe how naïve we were then, cocooned in our whiteness. But I was about to find out how ingrained racism was in the Navy then. More about that later.

The Treasure Island Navy base took up the entire island, a flat, manmade extension of the natural island of Yerba Buena, which anchors the Bay Bridge halfway between Oakland and San Francisco. The whole surface of Treasure Island was only a few feet above the water. I felt as if I were on the deck of a ship that had stopped sinking just in time. I don't remember signing in when I got there, though I'm sure it took forever, my first taste of military *hurry up and wait*. I do remember getting my uniforms. They were not the iconic sailor suit: a white hat, a neckerchief and blouse, and bell-bottom trousers with a thirteen-button fly, one for each of the thirteen original states. Officer candidate uniforms were much more mundane: khaki shirt and trousers, black socks, and shoes—and, of course, a necktie in dark navy blue to differentiate us from enlisted men, as clear a class signal as the difference between the businessman's suit and the laborer's jeans and hoodie.

I'm six feet, five and a half inches. The shirts and trousers the sailor slid across the counter to me, while looking me straight in the eye, were too small even for a much smaller person. "These won't fit," I said, pushing them back.

He pushed them right back to me. "You're right, they won't."

I pointed to the shelves behind him. "There's got to be some bigger ones there."

"Yup. There are."

"Well?"

"I'm saving them."

"For whom?"

He shrugged. "I don't know." Then, looking past me. "Next."

I was assigned a lower bunk in one of the barracks, long wooden buildings painted yellow. There was a row of lower and upper bunks on each side, leaving a passage down the center. The head (Navy talk for the bathroom) was at one end. We changed out of our civilian clothes into our uniforms, and I was a gawky, too tall teenager all over again whose arms were half a foot longer than his sleeves, making my large hands freakishly gigantic. My trouser cuffs just made it to the top of my socks so that my feet looked grotesquely long, as if I were standing in a pair of canoes. Had I appeared on stage in a comedy, I would only have to stand still to get laughs.

But my new barracks mates didn't laugh. My ridiculous-looking uniform was a signal to them of the contempt many enlisted men had for people, almost always younger than they, who got to be officers the easy way. Unlike new ensigns who, as midshipmen, had spent four arduous years at the Naval Academy, or the enlisted men who had worked their way up the ranks, we were going to get our commission in just two summer vacations, a total of ninety days of training, an opportunity withheld from people without a college degree. Along with our commissions, we would have a brand: *ninety-day wonder*, a term of derision. We all knew that there was nothing in our college educations that made us qualified to be candidates for commissioned officers. What was I going to do with my BA in literature? Gather the sailors around me just before taps and read Faulkner to them as they fell asleep?

After changing into our uniforms, we marched to the huge mess hall for noontime chow. Hundreds of men were eating at long tables. Yards away, some sailors were eating standing up—Marines, in their khaki. They were guarding the men at the tables. "From the brig," someone explained. "They get five minutes, that's all."

We got in line, approaching the end of one of several long counters where those already addicted picked up as many sample packages of Chesterfield cigarettes as their pockets could hold, and others, not smokers yet, took them just because they were free. Then we picked up our metal trays, which were divided into compartments, and held them out across the counter to the messmen, whose animosity radiated across the counter along with the food as they filled the trays: meat in one compartment, vegetables in others, and dessert in another. Rabbit was served that day. Fried, like chicken. I'd never even contemplated eating a rabbit before, cute little bunnies. It was delicious.

We could eat as much as the tray could hold, or until the few minutes allocated for chow were up. Then we carried our trays through the scullery, where we dropped them into slots in the washers, like dropping bread for giants into gigantic toasters. The air was full of steam and there were inches of water on the deck, which is what we were told to call the floor. Later we saw the sailors who were imprisoned in the brig crawl on their stomachs through the slimy water per orders barked by grinning Marines. I hadn't learned yet why Marines hated sailors and vice versa.

Some days later, in the mess hall for evening chow, the messman across the counter didn't place the scoop of mashed potatoes into one of the compartments on the tray I held out to him. He *threw* it. A good portion of it splattered off the tray and landed on my shirt and tie. I was halfway across the counter, ready to punch him out, before the guys to my right and left grabbed me by the arm and held me back. They were learning, faster than I, what was required to succeed in Reserve Officer Candidate School, otherwise known as ROC School: be passive when insulted or treated unfairly. And, of course, you had to pass the courses: navigation, ballistics, and naval orientation. Otherwise, say goodbye to college. Your Navy hitch began right then as an enlisted man, a white hat.

Which, I suppose, is why almost everyone cheated on exams. The stakes were too high for such niceties as personal integrity. What could be more unfathomable than the distance between college life and shipboard life in a war? I didn't cheat, though; I'm confident that none of the other of the twelve of us from Oberlin did. We'd been acculturated by Oberlin's honor system, which required the professor to leave the classroom when students were taking their tests.

There were some unique characters in our courses. At the very beginning of our first navigation class, the teacher wrote his name on the blackboard while we were still standing at attention. *Lieutenant Pancake.* There were a few muffled snickers. Then he turned around. "Gentlemen, you may laugh now." He held up his index finger. "Once when I give the signal. After that, if I hear the slightest murmur of a giggle, I'll flunk your asses. OK?"

No answer.

"OK. You ready?"

Still no answer.

"I'll take that for a yes. Here we go. One. Two. Three. NOW!"

Not a single giggle.

He nodded his head to show us he wasn't surprised at our cowardice. "Sit down."

I have no memory of the teacher of naval orientation, nor of ballistics. But I liked Mr. Pancake, even though I was incapable of keeping up with his blistering pace. One day, at the end of a class, he said, "Today I have covered what I would usually spend a week teaching the midshipmen at the Naval Academy. Guess whom I would trust more to find a continent, you guys or them?"

He was sure the Korean War would go on for a long time and we would fight in it. And then there would be another war and "they'll call you back in and you'll be fighting that one too. So, you better learn how." I didn't sense that he took any pleasure in his pronouncement.

* * * * *

WE GOT LIBERTY, OR TIME off, from 1700 hours (5 pm) on Friday until 0530 (5:30 am) on Saturday, when, as on every morning, we had fifteen minutes to get dressed, shaved, and our bunks made up before falling out into the early morning fog for inspections. Inspections always seemed to last longer than on weekdays, particularly for those candidates who'd stayed out all Friday night. Once, a candidate in front of me passed out, collapsing to the deck. He was lucky that this happened after his uniform had been inspected while he still managed to stand straight and tall at attention. The inspection went on as if nothing had happened, while he remained prone on the cement. If you were found to have the slightest error on Saturday morning inspection—your tie not two-blocked precisely, your shoes not gleaming, a belt that missed a loop—you were confined to base when liberty started again at noon, lasting until 2000 (8 pm) on Sunday. While your friends were happily riding on the Key System's light rail train, which ran in the center of the lower level of the Bay Bridge heading for San Francisco, you commenced the liberty period on the grinder, an area covered in tired blacktop where you marched double time to nowhere for several hours, carrying a World War I-era rifle with a cork stuffed into the business end, just to make sure you felt as ridiculous as you looked.

We all fell in love with San Francisco, of course. No one ever asked us if we were old enough to drink. Everyone thought we were about to be shipped out to Korea. At Tarantino's, a favorite restaurant of tourists on Fishermen's Wharf, I ordered a hamburger and was brought a steak. No one seemed to notice how gawky I looked in my too small uniform. Sometimes on Friday nights, we'd return to base just long enough to grab a few hours of sleep before inspection and leave again right after. Saturday nights we'd stay at the YMCA on the Embarcadero, near the Ferry Building. The transition from heaven during liberty to hell on Sunday evening was as dispiriting as it was abrupt. Next Friday evening would feel a century away.

The only enjoyment, aside from the ample food, was the hour and a half of sports every afternoon. The Navy brass operated under the illusion, reverently held, that athletes make the best leaders. Therefore, participation, either in basketball or flag football was required. I chose flag football. After a few days, the chief petty officers in charge separated the wheat from the chaff. So, along with the others from Oberlin, I played with or against the best. We had the small college athlete's chip on the shoulder, and so I reveled in the discovery that I could get free of a defender from Michigan State or Notre Dame and catch the pass. I still do, all these decades later. Absurd, I know, but so what? Once, right after dinner, clad in our heavy black shoes and uniforms, another candidate who was on the track team of a big state university and I raced each other the quarter of a mile or so from the mess hall to the barracks. He won.

But he threw up and I didn't.

Our company commanders were chief petty officers, the highest non-commissioned rank. Their assistants were officer candidates in their second summer. I don't know how these assistants were chosen. Our company's assistant commander had just graduated from college; as soon as ROC School completed in late August, he would be commissioned. He was very short, and I sensed early on he hated me for being tall. At morning inspection, he would spend more time on me than on anyone else, eagerly looking for some flaw in my uniform or dullness in the gleam of my shoes.

One day, while I stood stiffly at attention, he stepped on my right shoe, newly polished, and moved his foot around, destroying the shine, all the while looking up at me. "Your shoe is unshined," he said. I refused to look at him, which was easy. All I had to do was look straight ahead. I felt a rush of enjoyable hate. I'm sure he sensed it. Maybe it made him feel good too. "You get the grinder," he told me. He moved to my right and spent much less time inspecting the candidate next to me.

That Saturday, about twenty of us, miscreants all, stood at attention on the blacktop, ready to start jogging at the command of a pissed-off chief petty officer who'd caught the duty. He was going to take it out on us. And take it out he did—on everyone except me. Most of the candidates were staggering long before the two hours were over. Some fell to their knees, the pace of others reduced to a shamble. Not me. I could have run until Christmas and not breathed hard. I never slowed down, never stopped grinning, and every time I went by that pissed-off CPO, I winked at him. Some officer and gentleman! I'd never felt more victorious.

A few days after my adventure on the Grinder during a recess from navigation class, I sidled up to the assistant company commander, out of everyone's earshot. "You better stay on base during liberty times," I said. "Because if you go off base, I'll follow you and kill you." He went white in the face. I think he actually believed me. I never did check to see when liberty started if he stayed on base. I'd already had my revenge when his face went white, and I was sure somebody who had been on the grinder with me had told him how much I had enjoyed it. Looking back on this now, reexperiencing it by writing it, I feel the same childish glee I felt then. So much for aging into wisdom!

* * * * *

THE NEXT SUMMER, ROC SCHOOL was based in Southern California at the Long Beach Naval Station, now defunct. As soon as Saturday liberty started, we'd head for a beach to go body surfing. I had already fallen in love with California via the charms of San Francisco. Now, riding bigger, more powerful waves than existed on the East Coast, under the cloudless blue Southern California sky, I fell even deeper.

We'd swim out and tread water, waiting for the right wave. They came at us like hissing mountains. When we saw one with a steep side, we'd swim in front of it as fast as we could. Then it would be under us, lifting; if we hadn't timed it right, it would drop us again, rolling away under us, and we'd see the back of it like a long, silent blue hill sliding away, big enough to obscure the beach so that you felt as if you were miles and miles from land. Then we'd wait again. When we caught a wave just right, we'd surrender to its power. Up on its crest we could see the beach, people wading in the shallow, looking like little dolls, then we'd be lifted higher as the wave grew steeper, and for an instant we could look down at the beach ahead, like a pilot coming in for a landing—then we would be hurled down and then in foam, our arms straight out, and we'd land chest first in the sand before we could stagger to our feet.

Sometimes, instead, we'd get too far ahead of the wave so that it would flip us and then crash down on us, rolling us over and over, and we'd somersault and bounce on the sand, tucking our chins on our chests so we wouldn't break our necks. We'd hold our breaths under that enormous weight, until suddenly the wave would recede and leave us on the beach, uninjured and foolishly assuming we wouldn't be the next time either.

We'd turn around, facing the ocean, catch our breath, and wade back in, diving under each wave, one after another, until our feet didn't touch the sand anymore, and then swim out to wait for another ride and do it all over again.

Once, while treading water waiting for a wave, I saw my friend and teammate Jeff Blodgett a few yards to my right. I swam halfway to him, then dove and swam underwater to where, through the gloom, I saw his legs treading water, as if he were riding a bike. I grabbed his ankles and pulled him under and held him there, trying not to giggle and choke, so proud of myself for thinking of doing this to him before he did it to me. I let him go after a few seconds and bobbed to the surface to discover he wasn't Jeff after all but a perfect stranger, terrified and gasping for breath, sure I was intent on drowning him just for the hell of it. I made an executive decision to depart.

I swam as fast as I could away from him parallel to the beach and then caught the best wave of the day. It projected me toward the shore, right at a mountain of suntanned flesh, a woman who had been placidly cooling her ankles in the thin foam the waves left behind, until she saw this very large person hurtling toward her like a torpedo. I collided with her in a tangle of arms and legs and hirsute armpits flattening her on the sand. Then I jumped up and quickly departed, running down the beach under a small airplane that towed a banner informing the world that JESUS SAVES. Nothing like that combination of events ever happened in Riverside, Connecticut, where I had grown up. I'd have been in heaven, if I hadn't been so lonely for Joanna,

My father had five siblings, all older than he. His eldest sister, my aunt, Isabel Shanklin, had married a Californian. She had lost her husband but still lived in their big house on Wilshire Boulevard in LA. The Shanklin family had a second home in Hermosa Beach only a few blocks from the surf. On several Saturdays, they invited my Oberlin ROC School friends and me to lunch, which we ate outdoors on the patio in the bugless California air. Then we all would head for the beach, where it seemed that everyone was young, tanned, long muscled, and impossibly healthy, avatars of the life the Beach Boys would sing about decades later.

Aunt Isabel's daughter, my cousin Edith, who would become a lifelong friend, was a teacher at a private girls' school at the time. One night early in our stay, she invited some seniors and recent graduates to my aunt's capacious house for dinner with us. I believe all twelve of us from Oberlin were there that night, with the same number of girls. We ate outdoors on the back patio in the balmy air and the smell of jasmine, and then we jumped in huge, open shiny convertibles with white tires and drove through streets which I remember as being less crowded than I am sure they were. We went to the Lighthouse Café in Hermosa Beach, where we listened to jazz and drank fizzy, gin-infused drinks, and then walked on the beach next to the thin foam the waves left behind, white against the black roaring ocean.

I was driven back to my aunt's house that night. I don't know where the eleven others slept. Most probably back at the base, but I like to think that at least some of them slept on the beach. All they'd have to do in the morning to get clean for the day would be to take a dip in the ocean.

In the morning I went to church with my Aunt Isabel. I pretended that I, too, wanted to go to church, though what I really wanted to do was get to the beach as soon as I could. That afternoon, and almost all the following Sunday afternoons, the same group of boys and girls reconvened on a beach. I remember every one of those girls as impossibly beautiful, hard not to stare at in their bathing suits. How carefree they seemed! They'd always drive us back to the naval base just in time. We'd walk through the gates into yet another world.

* * * * *

I DON'T HAVE EVEN THE foggiest memory of what I learned in ballistics that second summer. All I remember from naval orientation was something about leaving a calling card in a tray near the front door of a senior officer and his wife when calling on them, which I had no intention of ever doing. I do remember how dumb I was in navigation, afraid of failing the course and thus being put aboard a ship somewhere as an enlisted man instead of returning to Oberlin for senior year. The night before the final exam, I got in my bunk, as required at taps. A few minutes later, I went to the head with the textbook and sat on a toilet studying until late into the night. Nevertheless, when I finished plotting a long cruise in a fictional destroyer the next morning as the main part of the exam, I ended up on dry land, somewhere between the Pacific Ocean and the Sierra Nevada Mountains. Hoping that the officer who would grade the exam had poor judgment and a lively sense of humor, I wrote at the bottom: *This vessel is amphibious.* And then waited and waited and waited for the bad news.

It never came. I figured the Navy had spent a lot of money they didn't want to waste feeding me, buying my ill-fitting uniforms, and transporting me across the country twice. Unlike on merchant marine ships where each watch officer navigates during his watch on the bridge, the Navy appoints a navigator to do all the navigation and not stand bridge watches. Obviously, that navigator would not be me.

In May of 1953, less than a month before we graduated from Oberlin, the Korean War ended. So, just as we'd been too young to have to fight in WWII, we were saved again. We'd be playing at war instead of fighting in a real one. By the time the Vietnam War would start, we'd be too old. My military experience was about to be very different from that of my cousins, my dad's older brothers' sons who were at most ten years older than I. They experienced horrors that were literally unspeakable— except for my cousin Roland's observation about a funeral he had attended shortly after his demobilization: *How strange it seems to hold a funeral for only one person at a time.*

Suppose that instead of sending young people off to die, nations all over the world agreed that no citizen would be drafted into the military until they were ninety years old, my age now; nor would anyone younger than ninety be allowed to volunteer their services to the military. Do you think we would have any wars?

CHAPTER TWO

My two-year hitch of active duty wouldn't begin until August 5th. I found a job for the time until then, laboring for a Marine contractor in Stamford, Connecticut, not far from my family's home in Riverside. A former Seabee enlisted man, the contractor treated me courteously and fairly, commending me for working hard, wishing me luck in the Navy—until I carelessly let out that I wasn't a white hat but a ninety-day wonder. Immediately, his attitude changed. I wasn't "Steve" anymore. I was "Ensign Davenport," spoken with contempt. It didn't bother me much. I had learned in the second summer of ROC School to accept the enlisted men's animosity for the easy and privileged way we got to be their bosses.

We were extending a dock, making it reach farther out into Long Island Sound. It required taking loads of cement in wheelbarrows from a cement truck and wheeling them out to the end of the dock, then dumping the cement into cylindered forms to make the supports that would hold up the dock. I was more challenged by boredom than fatigue. I liked the feeling of heavy work. I liked to sweat, and I wanted to impress the contractor, a guy who, unlike all the men I knew in their funeral business suits, worked in jeans and T-shirt or no shirt at all in the humid air of summer. He'd made his own business, was his own boss. My hard work didn't change his attitude, though. Looking back now, I figure he didn't find anything impressive about a person who felt the need to impress him.

"Ensign Davenport, I've got special job for you," he said just as I showed up for work one morning several weeks after he discovered that I was a ninety-day wonder. "Follow me." I was sure that whatever he had in store for me was going to be my punishment. He'd been casting around for ideas.

He led me down off the dock to the water's edge where an ancient barge was propped up on wooden horses in the mud. It was low tide. At high tide the water would lap against the legs of the horses, which seemed to be almost as rotted as the barge. I don't know why it was there, why it hadn't been broken up or sunk. The bottom of the barge was about level with my belt. He handed me a heavy pin maul with a long wooden handle and some spikes and told me to hammer the spikes upward into the rotten bottom. I knew my arms would be jelly after a dozen or so swings upward. I wouldn't be able to defeat him the way I had defeated the pissed-off chief petty officer on the grinder the summer before. And yet it didn't occur to me to tell him no and quit the job. The contractor sat on a big rock next to the barge to watch and taunt.

First, I had to lie down on my back in the mud below the barge and seat the spikes, getting them into the wooden beams of the bottom far enough so they wouldn't fall out. I held the pin maul with my right hand near the hammer, and with my left held the spike against the wood and tapped it into the barnacle-encrusted wood. But there was no way I could hammer them all the way in while lying on my back. I crawled out from under the barge and stood up, and swung blindly upward to where I thought the spike was partly embedded. I swung upward and missed twice, the hammer thudding into the wood. On the third try I hit the spike and it went a little way in.

"ONE!" the contractor said.

I swung, missing this time. "TWO!" the contractor said.

I swung again, missing again. "THREE!" the contractor said. This went on until he'd got to ten or eleven. I could tell I was going to be able to swing the pin maul upward only a few times more. I was pouring sweat, my arms were leaden, mosquitoes were swarming around me. I turned to face the contractor. He was still sitting down on the big rock. I lifted the pin maul above my head, my elbows pointed to the sky. I focused on the top of his head. I had no intention, whatsoever, of killing him. But if he flinched, I would win.

He didn't move a muscle, his eyes on mine, his expression didn't change. I put the maul down in the mud and walked away.

The next day, I carried a six-by-six beam, holding it halfway along its length, from a truck that had delivered it to where it was needed near the end of the dock. Passing by the contractor who was talking to somebody, I swung the beam so the end of it hit him on his butt. Not hard, exactly, but not easy either. I didn't make a conscious decision to do this. I simply obeyed an impulse. As soon as I had done it, I expected him to fire me, right then and there, or maybe even charge at me, like an angry grizzly bear. Instead, he turned from the person he was talking with, gave me a half smile, and nodded his head in approval, and the rest of my time working for him he treated me as he had before he'd learned that I was a ninety-day wonder. I left his employ in the middle of the last week of July. He paid me for all that week.

It occurs to me now that if I had needed the job as I'm sure his other employees did, I would never have had the impulse to bang him on his butt with a six-by-six—and if I'd had such an impulse, I certainly would not have obeyed it.

CHAPTER THREE

On August 4, I boarded a train headed south for the Little Creek Naval Amphibious Base near Norfolk, Virginia. I'd never been farther south on the East Coast than Philadelphia.

I changed trains in Richmond. I was wearing my brand-new uniform. (Its visored hat is right beside me as I write, memorabilia sitting on top of a lampshade.) Once at the base, I headed for a lunch counter. Behind it, a very pretty Black woman, about my age, in a black uniform dress and a white apron, watched me coming. She also wore a worried frown, but I didn't catch on. I was hungry, looking forward to the hamburger I would ask her for, rare, with a slice of tomato and a slice of onion on a toasted bun. I sat down, still not catching on. "You can't sit here," she said, so quietly only I could hear.

"What?"

She didn't answer. Clearly embarrassed. Maybe frightened?

"Oh!"

She nodded. She wanted me gone. And fast!

"Sorry," I said, and left.

I don't remember anything else about that trip. Not whether I sat down at the Whites Only counter and ordered a hamburger, nothing about the train, not even how I got from the Norfolk, Virginia, railroad station to the Little Creek Naval Base.

* * * * *

MY ORDERS WERE TO REPORT to the base for two weeks of training in amphibious landings: how to lead a wave of landing craft from ships anchored offshore to the assigned section of a beach at exactly the assigned time. After the training, I would report on board the *Vermillion*, AKA 107. Translation: Auxiliary Cargo Attack. For reasons which I am confident absolutely no one knows, and maybe no one ever did, the K is for the C in "Cargo." She was a freighter fitted with her own stevedoring equipment—huge booms and more running rigging than any clipper ship—for lifting the landing craft off their cradles on the main deck to the outboard side of the ship where their crews embarked in them, lowering them down into the water, and then loading them with the vehicles and equipment the landing crafts would carry to the beach. The Vermilion had three holds forward of the superstructure, which we called the house, and two aft, and carried twenty-four landing craft, five of which were LCMs (landing craft mechanized). We called them Mike Boats. They weighed fifty tons and were fifty feet long for carrying vehicles. There were also seventeen LCPs (landing craft personnel, known as Peter boats, or sometimes just P-boats) twenty-four feet long for carrying soldiers or Marines, landlubbers who, if they were going to be killed in a real war, would probably die on a beach. The *Vermillion* also carried the captain's gig, his own personal boat for going ashore.

My assignment was as First Division officer, in charge of all the space forward of the house. I would be giving orders to a chief petty officer who was expert in all the functions of the division from having actually performed them as he rose through the ranks. He would see that my orders were carried out.

During my two weeks at Little Creek, I lived in the bachelor officers' quarters, the BOQ in Navy parlance. I had a room all to myself. I was too lonely for Joanna to spend any time alone in it. Instead, I spent the evenings in the Officer's Club. It was housed in a building I dimly remember as an oceanside version of Tara where Scarlet O'Hara lived. Watching officers eat dinner with their wives didn't help my loneliness. Nor did the after-dinner, alcohol-fueled singalongs that went late into the night. Maybe if I could sing on key, I would have enjoyed them. But as evening wore on, I would feel more and more hollow. I would have been less lonely reading a good book in my single room.

I was immediately recruited to play on the Little Creek Naval Base football team. If I accepted the invitation, I would not report on board the Vermilion until the football season was finished, sometime around Thanksgiving. I knew that was wrong. Why should I get paid for playing a sport I loved while others had to go to sea? It was crazy. And my father's admonition over the phone when I told him was, "Don't. You have taken enough chances. You could get hurt." I didn't say anything. He wouldn't understand that when people want to do things that somebody always says is dangerous, like riding motorcycles, skiing fast down steep mountains, scuba diving, running rapids, rock climbing, or mountain biking, they never think about getting hurt. If you thought about it too much, you might not do it, and you want to do it. And afterward, when it's all over and you can't do it anymore, you're glad you did do it.

After the first week of training at Little Creek and I was still deciding whether to play, I went on board the *Vermillion* on a Saturday afternoon to say hello. On the ride out to where she was anchored, a supply officer told me that everyone on the ship loved the skipper, Captain Oliver G. Kirk, a bona fide war hero—and what a difference it made to serve under such a man. "The sailors love him because they know how much he cares for them. When some dickhead admiral orders him to keep the whole crew on board on a Saturday morning when we're in port just so he can inspect us, Captain Kirk just plain disobeys, just keeps the duty section aboard. I don't know how he gets away with it, but he does. The only time he gets angry is when somebody disobeys a safety regulation. When that happens, you don't want to be the officer in charge."

I told the supply officer about the football decision I was facing. He shook his head. "You think they'll send a substitute on board? There's gonna be just as many watches to stand, and one fewer officer to stand them. How do you think the other watch officers will feel about you when the football season's over and you finally do show up? A ship's not a good place to be hated. You can't go home, you're already there."

I decided right then not to play. That afternoon, I was toured around the ship by the gunnery officer, Ensign Nelson B., who would become a good friend. He told me we were going to go to the Med. We'd be getting liberty in ports in Italy, Spain, maybe Southern France! That news softened the disappointment that I'd never play football again.

The training at Little Creek finished, I reported aboard the *Vermillion* the next Saturday. On the way out to where she was anchored, a chief petty officer, the only other person on board with me besides the three-man crew of the Mike Boat, lit a cigarette. It was, and probably still is, against naval regulations to smoke on boats, or any vessel small enough to be carried on a ship. I was standing on the starboard side of the small afterdeck of the Mike Boat, holding on against the roll of the boat in the very moderate swell to the coxswain's cockpit, which was a square of steel coming up to the coxswain's shoulders designed to protect him from incoming bullets. The chief stood on the port side, just a foot or two from the stern, not holding on to anything. I didn't think he was smoking to challenge my authority. He wasn't even looking at me. He was smoking because he just wanted a smoke. But it was against regulations and as a commissioned officer I was the senior person. I was twenty-two; he was in his mid-forties at least, had served long enough to work his way up the ranks to be a chief. Most likely he'd survived combat at sea in WWII and maybe also in the Korean War.

I wasn't even on board yet and here I was, already having to assert my unearned authority on a person who was more qualified than I even if he were dead drunk, on drugs, sound asleep, and deathly ill, all at once. I looked the other way.

But what if he was smoking just to see if I had the nerve? I was terrified. The *Vermillion* was looking even bigger now. We were halfway there. I turned toward the chief. "Chief, put your cigarette out," I said. But not loud enough over the roar of the engines. Two of them, two hundred and fifty horsepower each, they made a lot of noise. The chief was looking straight ahead; he seemed planted to the rolling deck. "Chief," I yelled, "put it out!"

He looked across the deck at me, as if just then discovering my presence. I pointed to his cigarette between two fingers of his right hand and mimed throwing it overboard.

He looked surprised, uncomprehending, nothing challenging at all in his expression. The coxswain, encased in his cockpit up to his shoulders, turned to look at me, waiting for my next move. He'd been close enough to hear. I wanted to put him in his place by telling him to keep his eyes on where he was steering, but I didn't have the balls.

The chief formed the word with his lips. Slowly. Clearly. I couldn't mistake it. "Diesel."

"Diesel?" I mouthed.

He shook his head up and down. Vigorously. Wanting me to understand. The white foaming wake was just behind and a little below his ankles.

"Chief! Throw it overboard," I yelled.

He looked disappointed. Like a teacher staring at the dumbest boy in class. He stepped forward, away from the brink at last, bent down, and opened the cap of the fuel tank. Then he stood up straight holding the cigarette above his head, his wrist bent down, aiming. And dropped the cigarette in. He carefully screwed the cap back on, not once glancing my way. He was sure by then that I knew it was safe to smoke on Mike Boats, powered not by gasoline but diesel oil.

The coxswain maneuvered the Mike Boat up to the platform at the bottom of the *Vermillion*'s gangway—something I would learn was not easy to do. I climbed the gangway, stepped on to the main deck, and saluted the flag at the stern, also called the ensign, then I saluted the officer of the deck. "Request permission to come aboard," I said.

"Granted, and welcome aboard." The officer of the deck, an ensign too, didn't look anywhere near old enough to have graduated from high school, let alone college. I soon learned he was from Florida and that the enlisted men referred to him as Mr. Hushpuppy—after some weird Floridian food he told everybody he liked to eat. I was even less experienced than he, probably just as incompetent, so he was glad to meet me.

Someone led me to my quarters, a tiny stateroom on the second level of the house, shared with three other ensigns. There was an upper and lower bunk on each side of the room, with a space in the middle just big enough to stand in. Nobody was there when I arrived. I felt an explosion of the loneliness I'd felt at the BOQ. Everything was gray.

After the incident on the way to the ship, I was even more aware how much my success would depend on a good relationship with the chief petty officer in charge of the First Division. My instinct told me I should find him and introduce myself right away. But I had barely begun to unpack my stuff when he poked his head into the stateroom. "Welcome aboard, sir," he said, sticking out his hand. "I'm Chief W." He looked about fifty years old, a short, powerfully built man with a round belly. He carried his billed cap in his right hand. His crew cut was gray.

"Thanks, chief," I said, crestfallen, my plan already thwarted. "I'm Ensign Davenport."

He nodded, letting me know he already knew my name. "I need you on the deck, sir," he said.

Already? I thought. *I've only been on board a minute.* But I held my tongue and followed him down the long skinny passageway to an outboard ladder where he put his hat back on. He led me down to the main deck and then forward to the space between the house and the bow, where three of the ship's five holds were located. I was technically in charge of this space as the First Division officer, the chief's boss. I was petrified he was going to ask me for advice about some problem he was facing.

It was a large, very gray space. The ship was quiet; two-thirds of the crew and officers were ashore on weekend liberty. Above us loomed two huge Mike Boats on hatch number three. A P-boat was nestled in each of them. Further toward the bow, it was the same on hatch two: two Mike Boats, each containing a P-boat.

"I'm worried about this starboard list," the chief said.

"What list? I don't see a list."

"You don't?" He looked surprised.

"Well, I guess I do, yeah, I do. Definitely a list."

He nodded. "Good. Sometimes it's hard to notice it. You know? When you first get aboard?" The chief took off his billed cap, scratched his gray hair, and put the hat back on. I knew right then I was going to be witness to this gesture a gajillion times. "So, what do you think we should do about it?" he asked.

"Geez, chief. I haven't the foggiest idea." What else could I say? Besides, he seemed like a nice guy. Old enough to be my father.

"I have an idea," he said.

"What?"

"The boats are all misaligned."

"Misaligned?"

"Yes, sir. Misaligned."

"So?"

He pursed his lips and nodded. "That's what we're gonna have to do, sir. I'll go get the men."

"You're gonna align them?"

"Yes, sir. We're going to align today. If you give the order, that is, sir."

"Aligning is something that happens a lot?"

"Well, I wouldn't say a lot, sir."

"Just once in a while?"

"Yes, sir," he said, nodding vigorously. "Just once in a while. But when you have to do it, you have to do it."

"All right, chief, if you say so."

"I do, sir. Unless you have a better idea."

"Oh no, chief. I'm sure you're right."

"Thank you, sir. What we'll try to do first is just move the P-boats from hatch two to hatch three, and the ones from hatch three to hatch two, because, as you know, sir, it will be a big deal to have to move the Mike Boats too, especially with only one-third of your division on board."

"Good idea, chief. Let's hope that'll do it."

"Aye aye, sir." He left me and walked to where officer of the deck Ensign Hushpuppy stood near the gangway. A few seconds later, I heard the boson's pipe's shrill whistle over the loudspeaker, and then, "NOW HEAR THIS: FIRST DIVISION DUTY SECTION REPORT TO HATCH THREE." Minutes later, about a dozen men appeared on deck, dressed in work uniforms: blue cambric shirts, bell-bottom jeans. Some wore their hats, some didn't. The chief obviously didn't care.

Then, while I watched, a sailor climbed up into a P-boat that was in one of the Mike Boats on hatch two. Another man worked the winches of a huge boom so that a gigantic block, or pulley, was positioned above the P-boat. The sailor in the P-boat put a hook resembling a huge capital J that hung from the block into a metal ring that was attached to four cables, each leading to a corner of the boat. Then he jumped out and the winchman maneuvered the boom so that the P-boat was lifted out of the Mike Boat over our heads. Sailors grabbed lines attached to the four corners of the boat to keep it from swinging out of control as the ship rolled in the mild swell. (Later, ten miles off beaches in the open ocean in much bigger swells, I would often see men holding those lines lifted right off their feet.) The winchman swung the boat so that it faced forward and then lowered it to where it was level with the main deck. The three-man boat crew removed the railing that was there to keep men from falling overboard in rough seas and stepped into the boat, and the winchman swung it a few feet away from the ship and lowered it into the water. Thirty feet below the main deck, it rode one swell and the ship another, and the huge block, which I would learn weighed one ton, swung dangerously even in that mild swell. The winchman manipulated a cable that was attached to a small hook in the ring to pull the ring upward, taking the weight off the big J hook, and one of the boat crew, known as the boathook, yanked it out of the ring, while also dodging the swaying block. And the coxswain drove the boat away from the ship.

After that process was repeated for the other P-boat on hatch two, the chief turned to me. "Damn!" he said, shaking his head. "Just what I feared. Fucking list. Exactly the same."

"Really?" Over his shoulder I saw the two P-boats going in circles fifty yards or so away. Retroactively, I think I saw the crews laughing. But I was way too much in over my head to figure out what was really happening.

The chief removed his billed cap and scratched his gray hair and put his hat back on. "Yeah, really, sir. Gonna have to move the Mike Boats too."

So, the morning went on and on. I was feeling useless just standing there. When they lifted the Mike Boat off hatch one and its huge fifty-ton bulk swung over our heads, I tried to help by grabbing one of the lines. The chief, who had stood to one side giving orders through the whole maneuver, put his hand on my elbow and tugged me gently away, shaking his head, as if telling me to mind my own business. I felt utterly useless, totally beside the point. When they lifted the Mike Boat off hatch two and put it on hatch one, I tried not to recognize an uneasy feeling that I was being had. That got harder to do as they lifted the two P-boats back up.

At last, the job was done. The boats that had been on hatch one were now on hatch two, the boats that had been on hatch two were on hatch one. But the men hung around. Some of them pretended they weren't watching me for my reaction. It seemed like everybody was trying not to laugh. "Well, sir," the chief said, speaking loudly enough for all the men to hear. "We did it. See the difference?" There wasn't any, of course. The list to starboard was exactly the same.

"Right, chief. It's much better," I said, working hard to grin. I knew enough to understand that to get even a modicum of their respect, I had better be a good sport.

That evening, at dinner in the wardroom, with the other on-duty officers, I was asked in several different ways by several different grinning new shipmates how my first day on board had gone. I was also informed that the *Vermillion* had a permanent three-degree list to starboard that no one had ever been able to explain.

CHAPTER FOUR

We never did go to the Mediterranean. Either the news had been incorrect, or the orders were changed. Instead, we were ordered to steam south to Moorhead City, North Carolina. There, we would take on a load of vehicles and the Marines who operated them. We would then join a task force headed to Vieques, a small island near Puerto Rico where the Navy and Marines used to practice amphibious landings, much to the harm of the people who lived there. (In 2003, as a result of many protests, the Navy ceased operations there and the island declared a superfund environmental cleanup site. Now there are hotels for vacationers on the same beaches we "invaded.")

On the first day out of Norfolk, Captain Kirk called us to general quarters for the first of many drills. I donned my helmet and life jacket and ran down the starboard main deck passageway as fast as I could to my battle station, the twenty-millimeter-caliber guns on the foc'sle. I kept hearing a strange sound: *pop*, then a tinkle, like glass shattering. Over and over, *pop, tinkle pop, pop, tinkle pop*, while the passageway kept getting darker. A chief petty officer standing in the doorway of his mess room, who'd survived several real battles and was in much less of a hurry than I to get to his station, gave me an appalled look as I ran by in the deepening gloom, and I finally realized my helmet was colliding with the lightbulbs above me, exploding them one by one.

The loading facility in Moorhead City could only service one ship at a time, and there was no place to anchor, so we waited our turn several miles out to sea, steaming in a square about ten miles long each side, in view of land. In those days, when sexism was almost as deeply embedded in Navy culture as racism, women were not allowed to serve on ships. Moreover, only officers were permitted to invite civilian guests aboard and only while ships were in port. But even though we were not in port, for the several days we steamed in a square, we had a female aboard, Captain Kirk's daughter. The captain's quarters, where he took his meals alone, were commodious, with plenty of room for a separate place to sleep. What I remember most vividly about her presence among us was not the surprise that our skipper would make no effort to hide his violation of Navy regulations from us, but instead the affection and respect the enlisted men especially accorded for this one young female among us, a reflection of their feelings for her father. She was in the last years of her teens, maybe a freshman in college. When she sat in a swivel chair, reserved for the captain only, on the port or starboard wing of the bridge, watching us at our work or reading, the quartermasters would bring her coffee. When she took a walk on the main deck, sailors would smile, say good morning, and tip their hats. No one ever used profanity in her presence.

We were surrounded by boats fishing for shrimp. The radio they used to communicate with each other was, for some reason, on the same frequency as the short-distance radio we used to talk between ships steaming close together in formation. The supply officer saw an opportunity: we'd buy the whole load from whichever shrimp boat could get to us first. I was the junior officer of the deck who sent the message. Captain Kirk, sitting in his chair on the starboard wing, thought it was a great idea. The officer of the deck signaled the engine room, "All engines stop," and the *Vermillion* glided for a mile or so, finally losing all way, dead in the water, rolling in the swell, while several shrimp boats headed for us racing each other. Why do I remember this moment more vividly than others of equal non-import? Was it the contrast between our gray, uniformed, rigid military world and the freedom of those civilian fishermen? I can still sense the helplessness that a ship must feel when she's not driving forward, and I can still smell the tang of fish, saltwater, and diesel fumes that rose to us from the smaller boat thirty feet below us as the fishermen loaded big casks of shrimp into the cargo net and watched them ascend to us, their faces upturned. For the next three or four days, Captain Kirk and his daughter in his quarters, the officers in the wardroom, and the enlisted men down below had delicious fresh shrimp for dinner, fried one day, boiled another, in gumbo another.

Then it was our turn to dock in Moorhead City and take on our cargo. Captain Kirk's daughter departed ship, leaving us lonely for her and for the better selves we'd called up to merit her presence among us. As if in substitute, we filled the holds with vehicles and their gas fumes and stored their Marine drivers and mechanics in very tight quarters down below, which, at the first rolls and pitches of the ship, they would befoul with their vomitous seasickness—as good a reason to hate them as there ever was.

On the first day out from Moorhead City, I made the required daily inspection of the three forward holds, immediately after my watch on the bridge had ended at 1600 (4 pm). It was critical to make sure the vehicles were secured to the deck by chains. If one got loose even in a moderate swell, it could cause a lot of damage. The smell of gas and the roll and pitch of the ship made these holds very unpleasant places to be, especially for hold one at the bow, where the ride up and down as the ship pitched was most extreme. I worked my way forward, inspecting holds two and three, and then climbed down into hold one, the only place in the world I have ever felt seasick. A Marine private stood at stiff attention at the foot of the ladder, saluting me. Taken aback by his formality, I almost failed to return the salute. We were riding up and up and up, then pausing, then down, faster, and faster diving, then pausing again, and then up again, over and over, while he showed me that all was secure. I told him to make sure to go up topside to get fresh air frequently and, as if showing him how, turned quickly away and climbed the ladder up to the main deck.

The next day at the same time, I made my inspection and again was saluted in hold one by the same Marine private. Again, he accompanied me while I made my inspection, and we went up and down, up and down, and I began to feel sick. This time, I asked him his name and where he was from, and he told me. It occurred to me that since there were at least a hundred enlisted Marines aboard, it was strange that he had to stand watch every day, and even stranger than that was that it was the same time of day. "How come?" I asked.

"Sir, I haven't been relieved yet."

"You haven't!"

"No, sir."

"You've been here all this time since—"

"Yes, sir."

"Why?"

"I guess the sergeant forgot."

"Forgot! You get yourself topside right now and get some fresh air, you hear." It was the first time I'd felt legitimate giving an order.

He shook his head. "Not til I'm relieved, sir."

"Up!" I said pointing to the ladder.

"Please, sir, don't."

"You'll get in trouble?"

"Yes, sir."

I went topside and ran back aft toward the vomitous area where the Marines were incarcerated to find the sergeant. Halfway there, I changed directions and headed for the more commodious space on the second level in the house where the Marine officers were bunked. There, I told a second lieutenant; before I'd finished, he sprinted away to find the sergeant in charge, to chew him out and tell him to rectify the situation. The Marine private had been in that hold for twenty-four hours! There wasn't one sailor in the First Division who, when his relief didn't show up at the appointed time, wouldn't have run to Chief W to ask, "Where the hell is my relief?"

* * * * *

LATER, AS WE NEARED VIEQUES, the officer in tactical command ordered each ship to follow him "using your own discretion" through a make-believe narrow channel that earlier had been swept clear by minesweepers of mines laid down by the "enemy." The safe channel was fictional: no mines had been laid down, but the coordinates had been announced and our navigator, Pete Smith, an experienced merchant marine officer doing his two-year required hitch in the Navy Reserve, had marked them on the chart. We were practicing avoiding mistakes that in real war could sink us. Captain Kirk was sitting in his chair on the starboard wing of the bridge. Pete Smith was standing by the table that held the charts. He stepped away from the table to tell the officer of the deck the course to take when the OTC gave the signal. Then he returned to his chart.

A few minutes later, EXECUTE came over the loudspeaker, and all the ships lined up one after another in the assigned order behind the ship commanded by the OTC, and followed him as he turned left to enter the channel.

Except the *Vermillion*. We turned *right*. Captain Kirk stayed in his chair. He didn't say anything.

Several minutes later, we heard over the loudspeaker, "Admiral X-ray, this is Hashfield Peter. Captain, will you join us?" We were sure it was the OTC's sardonic tone, not the junior officer of the deck whose job it was to guard the talk-between-ships radio.

Everybody on the bridge, even the helmsman glanced at Pete.

Captain Kirk still didn't get out of his chair, and he still didn't say anything. He just turned his head to look at Pete.

"He's wrong and we're right," Pete said. He unpinned the chart from the table and turned to carry it toward Captain Kirk, eager to show him.

Captain Kirk put up his hand, a stop sign.

Pete stopped in his tracks. "Thank you, sir!"

Captain Kirk nodded, then looked at me. "Mr. Davenport, tell him no."

I felt a surge of joy. "Hashfield Peter," I said gleefully into the transmitter, "this is Admiral X-ray. Reference your last transmission: NEGATIVE!"

Captain Kirk could not have been the only commanding officer who knew the OTC's mistake. But it wasn't war, no ships would be sunk, no sailors drowned or burned to death if their commanding officers dutifully followed their commander who would write their fitness reports. So why take the risk of "using their discretion" and showing him up? If after the operation, Captain Kirk was proved to be wrong, he would have been officially reprimanded for making it clear to us all, and to Pete especially, how thoroughly he trusted his navigator. But we knew that the incident would never be mentioned. The truth was right there on Pete's chart.

CHAPTER FIVE

I was learning fast that I had no natural talents as a bridge watch officer conning a vessel as unmaneuverable as an AKA in a formation of ships steaming a thousand yards apart, a significantly shorter distance than the radius of an AKA's turning circle. Unlike sleeker combat ships with propellers on both sides of the rudder, helpful in turning the ship so that one propeller could be backed while the other went forward, our lumbering AKA had only one propeller. Even at hard right or left rudder, she would continue straight ahead for what seemed like forever before she responded and began to turn. Our standard speed was only twelve knots, about fourteen land miles per hour, but even when reversing the engines at emergency full speed, it took at least a mile to stop. Think of driving an automobile at high speed surrounded in front, on both sides, and behind by other cars only a few yards away. Each ship was assigned a specific distance away and compass bearing from the guide ship, which had the easy task of always being in the right position. To stay on station required fighting a combination of boredom and tension while relentlessly checking and adjusting speed and direction. Even as the only junior officer of the deck, serving as an assistant to the officer of the deck, I was slow to figure things out, constantly on edge, faking a confidence I didn't feel.

We went to Vieques three times when I was aboard the *Vermillion*. We spent some of the time between those cruises taking Naval Reservists on their required annual two weeks of active duty, leaving port on Monday mornings, returning two Saturdays later. On one of those cruises, Captain Kirk gave Ensign Nelson B., Ensign "Hushpuppy" J., and me a lesson in ship handling.

We were standing on top of the house, a level above the bridge where the vision is best. The ship was steaming away from a huge red buoy that had been tossed overboard floating in the wake behind us. The object was to get back to it and come to a stop beside it as if we were maneuvering up to a dock without the aid of tugboats. "You go first, Mr. Davenport," Captain Kirk said. Maybe he thought we would decide to stay in the Navy when our required time was up. Maybe he was trying to entice us. I didn't know then that being a leader is being a teacher.

When the buoy was small in my binoculars, I judged that it was far enough behind us to provide room for the ship to circle back to it. I leaned over the speaking tube and told the helmsman below on the bridge to come left. He repeated the order and obeyed, but the ship did not. It seemed forever that she plowed straight ahead. When at last she responded, I gave another order and another, until we headed straight for the red buoy. I told the helmsman to aim for it and ordered the engine room to take a specific number of turns off, reducing the number of propeller revolutions per minute, and then a minute later to stop engines; but the ship continued plowing straight ahead.

"Reverse emergency full!" I yelled into the speaking tube, and the ship started to shake and shudder, and now neither the helmsman nor I could see the buoy over the gun tubs on the focs'le. And then I did see it, a half a football field to starboard, and we sailed right on by.

Hushpuppy did no better. Nelson was so seasick that he "docked" the *Vermillion* so far from the buoy, we could only see it in the binoculars. And then Captain Kirk showed us how, explaining all the while—until the ship was dead in the water, and we looked down at the buoy right beside us.

"What would you do if it were a man overboard instead of a buoy?" Captain Kirk asked.

We spoke all at once interrupting each other, even the nauseated Nelson, to show him we knew. "Hard turn to whichever side the victim fell from so the propeller won't chew him up, then keep turning until you've added sixty degrees if you're turning right, subtracting if left, then keep on until you're on a reciprocal course to your original."

I only served under Captain Kirk for a few months before he left us many man-overboard drills later for a new assignment, but all these years later he looms in my memory for the way he behaved as our leader. What an admirable person! Even though many of the crew and officers were Reservists, eagerly waiting for our required active duty to end, we wanted to excel for him. Every time in our practice invasions when the order came to get all the boats in the water, the Vermilion was always first.

At the end of his career, Captain Kirk was a rear admiral; but that's not how I think of him. I think of him as a sailor, a man of the sea, our leader who cared for those he led more than he cared for anything else.

* * * * *

WE ARE RETURNING TO PORT after two weeks of training Reservists on their required annual duty. It is midwinter, after dark. I'm standing at the bow in charge of the anchor detail. Captain Kirk is conning us toward our assigned anchorage, a mere dot, a tiny circle on the chart. There's a lot of traffic, ferry boats, tugboats, barges—I have no idea where we are in relation to our anchorage that on the water has no mark. After a while, the vibrating throb of the engines we never notice because we always hear it stops, and now the sound we hear is sibilant as we glide. A minute later, the throb begins again, the ship shuddering in reverse. And then we are dead in the water, exactly where we should be. In my earphones, I get the order: drop the starboard anchor, which I relay to sailors who, unlike me, know how to do it, and there's now the sound of the chain rattling and the big splash below. Minutes later, the First Division puts the gangway over the side and drops a Mike Boat for everybody who will go ashore on liberty. But not the captain's gig. He, long married, is taking the duty for the assigned command duty officer, a lieutenant (junior grade) who was married just days before we'd gone to sea.

That was our last cruise under Captain Kirk. He was transferred to some other, larger responsibility, and a new commanding officer came aboard. He was Captain James T. Lay, son-in-law of Fleet Admiral Chester Nimitz, who had been commander in chief of the US Pacific Fleet in World War II. But, to us, Captain Lay was what he wasn't—namely, Captain Kirk, to whom we instinctively compared him. The way he walked was different, the way he talked, how he wore his uniform—utterly meaningless signifiers to which we gave meaning without realizing how stupid we were being. It is hard to know whether he lacked an aura of leadership or whether we stripped his from him by refusing to acknowledge it. But Captain Lay soldiered on and did a good job. Looking back now, I can't identify a flaw in his leadership. He didn't demean himself by trying to be like his predecessor, didn't seek our approval. All our operations under his command were successful. And therefore, so was he.

Which was more than I can say for my performance.

I was officer of the watch in the Combat Information Center (CIC) on the four-to-eight watch one morning on the way to Vieques when we received an order to change position within the formation. I made a quick calculation on a maneuvering board, a printed form for solving problems of relative motion and speed between moving ships—calculations, not incidentally, that Captain Lay routinely made in his head without using a maneuvering board on occasions when he took the con. I recommended turning left to a specific compass point at standard speed. What I didn't do was remember that another ship was positioned behind us and to our left. It had been positioned there in plain sight for several days, a neighbor, and was glaringly obvious on the radar screen. Unlike me, the officer of the deck was competent, but he must have been having an off day or was simply exhausted. When EXECUTE came over the loudspeaker, he took my recommendation and ordered the helmsman to come left. Minutes later, just as I remembered the ship astern, I felt the ship shudder and knew the officer of the deck had ordered emergency reverse and hard right rudder. Then I heard the horn of the ship on our port quarter make the five short warning blasts reserved for exactly what was happening: collision at sea. I rushed to the bridge, which on the *Vermillion* was just a few feet forward of the CIC space.

Even though the helm was turned hard right, the *Vermillion* continued turning left for what seemed like hours while the reversed propeller caused everything to shudder and shake and we kept on going forward. The other ship turned left so that if we collided, it would be our port side against her starboard, a massive sideswipe with the potential of sinking both ships. Captain Lay appeared on the bridge half-dressed and took the con. But there was nothing left to be done. The two ships got so close to each other the water between frothed up in spray. Standing on the port wing of the bridge, I could have shaken hands with the skipper of the other ship standing on the starboard wing of his ship. Instead, we just stared at each other. Then the two ships stopped sliding toward each other, running parallel at first, then at last moving clearly away.

And then James T. Lay, the *Vermillion*'s new captain, took me aside and yelled at me—and yelled at me, and yelled at me some more. The error was mine, but the Navy would have held him responsible. It was his decision to assign an incompetent as CIC watch officer. I had come just a few yards away from destroying his career and had nothing to say. "I'm sorry" wouldn't cut it.

My incompetence made it hard to earn the respect of the men in the First Division. They simply tolerated me. All I had going for me was my obvious admiration of them for their skills. Like operating the winch of a gigantic boom so delicately as to place a truck that is almost as wide as a Mike Boat into Mike Boat thirty feet below, when both the Mike Boat and the ship are rolling in the waves and without killing any of the crew. The coxswain of a Mike Boat had a lever to his right to control the starboard engine and one to his left to control the port engine. He would push the levers forward to engage the forward gear and pull backward to engage the reverse. On top of each lever was a stirrup which the coxswain twisted to the right to increase power, to his left to decrease. He steered with a steering wheel. Three controls. Two hands. And when you are backing off a beach after making a landing, you better not get sideways to the surf.

The young Marine or Army infantry officer goes on patrol with his soldiers. He carries and uses a weapon, can do almost everything that his men do. But I gave orders to do things I couldn't do myself. That would have been acceptable to the men if they didn't know that I didn't perform my other duties well, as a bridge or CIC watch officer. But they did know.

Except for ordinary seaman, the lowest rank, there were only a few men in the First Division as young as I was. Unlike the more technical work—radio, radar, gunnery, or navigation—the hard, physical, abovedeck work that was dangerous even in peacetime was performed by deckhands. As the Navy grew more and more technological, deckhands advanced through the ranks slowly. A boatswain mate third class was likely to be as many as ten years older than the third-class level in the other departments. Unlike me, these guys knew their way around the Navy. I learned from them that there were two ways to replace any of the myriad parts necessary to operate. I could submit an order in triplicate to some mysterious supply department operated by people who had never even seen an ocean, and wait and wait, and often be refused. Or I could bribe Boatswain Mate Second Class Richard Grizmala to go ashore and steal what we needed from sources known only to him by recommending emergency leave for him. (First, because his father had just died; the next time because his mother had expired; the third time because all his children simultaneously; after that I think it was his aunt.) Thanks to Grizmala and my mendacity, we were always well equipped.

CHAPTER SIX

My guys knew how to have fun. For instance, a moment I will remember forever:

We were part of a big, amphibious operation on maneuvers off the coast of North Carolina. The object was to land troops and materials as in an invasion. On this morning, I was the commander of a wave of six Mike Boats ordered to land on a specific stretch of government-owned beach. My boat was in the middle of the wave, with two boats to my left and three to my right, each about twenty yards apart. They were all guiding on me. A sailor, whose name, believe it or not, was Sealover, was the bow hook in the boat to my right. He was a very funny man who could imitate the gestures, speech, and stride of every officer on the ship, including mine. In the back of my mind, I wondered how he was going to make us laugh today.

We were running north, parallel to the beach, toward the line of departure, which was marked by a destroyer escort anchored a mile from the beach. We had to arrive there within one minute of either side of the assigned time. When we did, we would turn ninety degrees to the left and head for the beach. It was wintertime, just above freezing, and everything was gray: the sky, the water, our boats, the destroyer escort ahead—except the brownish-yellow beach, to our left. It, and the boats to my left and right, disappeared every time my boat dropped down into the trough between the swells.

I was eager to ameliorate Captain Lay's negative opinion of me by showing him that I'm good at this, that I can tell the coxswain how fast or slow to go so that my wave arrives at both the line of departure and, later, the beach itself precisely at the right time. It's the one thing I was good at and I did not want to screw it up.

That's why I was doing this the Right Way, as opposed to the Navy Way, which says we must, we absolutely must, be guided over our radio by an officer in one of the ship's CIC—initials that stand for the Command Information Center, or "Christ, I'm Confused!" which is the most frequent condition. But with so many boats in the water, each a little dot on a radar screen, the CIC officer seldom had the foggiest idea which boat is which.

A radioman climbed down the embarkation net and jumped into the boat, and we waited for the radio to be lowered on a line. It descended toward us looking like a big, khaki-colored suitcase, and while it was still above our heads, I grabbed it and by mistake on purpose slammed it against the side of the ship, rendering it helpfully mute. We untied it from the line, the radioman put his earphones on and listened to nothing, and away we went.

I looked at my watch. We were going to hit the line of departure right on time, but when I looked to my right, the boat next to mine was lagging a half boat length behind. I swung my arm forward, signaling to speed up. Seaman First Class Sealover, squatting on the afterdeck, sent me a foolish-looking grin across the water. To him, this, like everything else, was a comedy. I waited to tell the coxswain of my boat to turn left and head for the beach until my boat was on the crest of a swell so the others could see us turning. We finished the turn; I was sure I lined up right, but now the lagging boat to my right was even further behind.

I waved to it again to speed up. "What the f— are you doing?" I yelled at the coxswain of the lagging boat, but of course he couldn't hear. He was looking straight ahead, pretending, I realized later, not to hear me. I put up my hands in disgust. Sealover grinned again, pointed at the coxswain, pointed to his own head, then shrugged. I gave up. His boat was going to be late, that's all there was to it. Damn!

I looked to my left; those two boats were lined up with mine exactly. The beach was only fifty yards or so away. I looked to my right again;

Sealover was reaching down into the well of his Mike Boat. He picked up a broom, and now we were in the surf, on the crest of a wave just before it would break, accelerating us toward the beach—and the boat to my right caught up, exactly aligned with mine!

Because Sealover was paddling madly with the broom.

* * * * *

WE WERE IN PORT. IT was still wintertime, just above freezing. We had the luxury of being tied up at a dock in the Norfolk Naval Base, much more convenient than being way out at anchor. I was the officer of the deck, standing by the gangway. It was about two am, halfway through the midwatch. I was tired and bored to death, and here came two of my guys, one tall and one short, walking down the dock toward the gangway. They were talking loudly, arguing about something. They'd been drinking, obviously. What else was there to do in Norfolk for sailors who weren't married besides picking up some woman in a bar, maybe getting laid, with or without paying, or getting in a fight?

"Welcome aboard," I said. "Did you have a good time?"

"Super duper," the tall one said.

The short one nodded, swaying a little, grinning. "I was going to bring you a beer," he said to me, "but"—pointing to his friend — "he wouldn't let me."

"Yeah, I told him you were too young," the tall one said. "You could get arrested."

I was giggling now. Just looking at them I wanted to laugh. I was a little jealous too. Per regulations, I was wearing a pistol in a holster at my belt, the magazine safely in my trouser pocket. Who the hell was *I* going to shoot?

"Nighty night," the short one said, and went below to where the enlisted men slept.

His friend crossed over to the outboard side, leaned on the rail, and gazed at the night.

A minute or two later, the short one reappeared. He was in his skivvies and a T-shirt, underneath which have blossomed two gigantic bouncing breasts made of his uniform. On his head was planted the business end of a mop. Its long, gray strands came down to his shoulders. He looked at me and put his forefinger against his lips. "Shhh." He was a hairy, limbed, seagoing seductress ready to play. He crossed to where his friend was still gazing peacefully at the night. "Hey, big boy, you want some of this," the seductress said, one hand beneath his boobs, jiggling, the other on his crotch.

The tall one turned, took one look, and—I kid you not—jumped overboard.

The boatswain mate on watch threw the life ring over, almost hitting him on the head. But the tall man calmly swam away from it, toward the stern and around to the dock where it was much lower and closer to the water. We threw another ring down to him and hauled him up.

"Geez, that was cold," he said, sounding surprised.

* * * * *

THE NEXT TIME WE WERE so lucky as to be tied up to a dock was in Portland, Maine. It was summertime and I had the midwatch again and here came a sailor on a horse. He was in his skivvies and T-shirt and a cowboy hat, the brim of which, propped up by his ears, hid most of his forehead. He calmly rode the horse up the gangway.

"Whoa," I yelled.

The sailor-cowboy pulled on the reins, and the horse stopped and plopped manure onto the gangway. His rider saluted, requesting permission to come aboard.

"Permission granted," I said, "but not him," pointing to the horse.

"It's a her," he corrected, dismounting. He guided the horse to turn around, patted her on the rear, and the horse, still fertilizing the gangway, sauntered down it onto the dock and out of sight, while her rider explained that he had met a very nice lady. In a funny accent he'd never heard before, she had told him how much she liked his Navy uniform. "Even the pants," he told me, still amazed by his good fortune. He'd warned her: *every time you want to take a leak, you got to undo thirteen buttons.* That made her like them even more! "Okay, let's trade," he'd suggested. "The horse and that hat for my Navy blues."

"Well," I said, "at least you get to keep the hat."

* * * * *

THE STATEROOM I SHARED WITH three other junior officers was tiny. The latch to the door of the head didn't work, so the slightest roll or pitch of the ship, even when anchored, caused the door to slam metal on metal, each harsh, ringing clang in tune with the motion of the ship. It was louder in rough seas than in calm, so we didn't perceive it as a sound—until we left the stateroom and heard the strange empty silence of everything else. My bunk, an upper, was shorter than I was. It had a solid wooden railing on the outer side to keep me from falling out when the ship rolled and was so close to the heating duct on the overhead that I couldn't sleep on my side. There was a porthole right beside my head. In the winter, I kept it open and slept on top of the blankets.

We were not allowed to make up our own bunks. That was the work of Black people, steward's mates. Obviously, America was much safer because I didn't have to make up my own bed. And when I sat down to eat in the wardroom, reserved for officers, a Black person waited on me, brought my food, took my orders. When I stood the midwatch or the four-to-eight in the morning, he was there also taking his watch, making sure my coffee was warm.

The *Vermillion*'s crew across all the departments was undermanned. The Second Division, whose responsibility aft of the house was the same as the First Division forward of the house, was the more undermanned of the two. So, when the officer in charge of the deck department, Lieutenant Lucky L., told Lieutenant Junior Grade Tom W., who was in charge of the Second Division, that he was assigning the boatswains' mate second class who had just come aboard to the Second Division, Tom was very pleased—until he learned that the man was Black.

"I don't want him," Tom said.

"I'll take him," I said.

"You got him," Lucky said. And that was that. The new boatswain's mate, unsurprisingly, turned out to be a high performer. He had to be.

A tall Black man named Washington preached to us when we were at sea on Sundays. In civilian life he was a pastor; in the Navy, he was a steward first class, too spiritually advanced to resent having to wait on white officers while being named after our most celebrated enslaver. Sometimes we would be so privileged as to be preached to by a real Navy chaplain, an officer—therefore, unlike Washington, officially a gentleman who arrived to us on a bosun's chair, precariously hanging from a line between two ships steaming dangerously close to one another. I admired those chaplains, commuting all day from one ship to another, but I was more inspired to be in Washington's aura, listening to his gentle admonitions to believe that in the end, all our sins would be redeemed. The services were held where the enlisted men slept and ate. On Sunday mornings, some of the men who were not on watch stayed in their bunks to catch up on sleep. So, Washington preached and led our prayers quietly as we moved up close to him so we could hear. He towered over us, a mixed group of enlisted and officers, looking up to him, catching his spirit. I felt a peacefulness that had never come to me in "real" church services. I knew it would leave me as soon as I went up topside, but while it lasted, it was a blessing.

* * * * *

AFTER ONE OF OUR OPERATIONS at Vieques was finished, we visited

St. Thomas for the Fourth of July weekend. While in port at Norfolk, I had racked up a good number of IOU's by taking the duty for married officers who had much more reason to go ashore than I. Now I would be free for the whole three-day liberty on this exotic Caribbean island.

The dock at the abandoned submarine base on St. Thomas was shorter than the *Vermillion*, and because there were no tugboats, we had to use our own Mike Boats to maneuver ourselves to it. This time we didn't break any radios because that's how I delivered the orders Captain Lay telephoned to me from the bridge to the coxswain in the Mike Boat pushing our bow, as did Third Division Officer Ensign Tom W. to the Mike Boat at our stern. It took forever, while the scents of lush vegetation and barbecuing meat wafted to us and our lust for freedom accelerated.

Coming down the gangway at last in the late afternoon, Ensign Hushpuppy and I were greeted on the dock by Caribbean women in beautiful colorful dresses, selling us in their delicious accents food we could eat in our hands. We were too excited to be hungry and suddenly seasick on this new surface that didn't move. As officers, we were permitted to wear civilian clothes when ashore—another form of liberation, not permitted to mere enlisted men. I was dressed in a new seersucker suit, a white button-down shirt and yellow tie, a preppy arriving where pirates once had.

The mother of a college friend owned a shop in town. I wanted to say hello to her and ask her where we could have the most fun, the best beaches, restaurants, whatever. Hushpuppy was married and I was engaged, so though intensely horny and not having seen, let alone talked to, a woman for over a month, we were not looking for romance. "Just company," Hushpuppy agreed. I had no way of knowing whether to believe him.

I don't remember what my friend's mother sold in her shop, but I do remember that as soon as Hushpuppy and I appeared, she closed the shop, pulled the blinds down, and we and her staff started to party. She was my parents' age; she could have passed for a Riverside matron, a staunch, corseted Republican hostess of mannered cocktail parties where no one ever sat down, but here she was overseeing a hilarious rum-enhanced celebration of the Fourth of July. By evening time, it devolved into a water-gun battle for no other reason than she thought it would be fun.

She also thought it would be fun to have dinner at a new hotel run by a friend who had recently emigrated from Bridgeport, Connecticut. "He has no idea how to run a hotel," she said. "It's utter chaos. In the restaurant, the guests help with the cooking. It's tremendous fun."

The hotel was on Water Island, literally an island in the harbor. We went out there in the hotel's boat, driven by a young Black man. All the windows of the hotel's dining room, which was also a dance floor, were open—or maybe there were no windows, just a roof. We were enfolded in dimmed shimmering light, lush tropic air, and the thin, satisfying rhythms of a steel band. I also remember wishing I could expend the energy expanding in me by making music like that to get lost in. My friend's mother strolled to the podium. The music stopped. She announced that she and her friends, Hushpuppy, and Steve of the United States Navy, so far from home, had come to celebrate the Fourth of July. A large family having dinner clapped their hands in applause and, switching from Spanish to English, invited us to join them. The steel band started up again.

The family was from San Juan, Puerto Rico. The patriarch was a doctor. There were also two sons, both slightly older than Hushpuppy and me, black hair, handsome, courtly, formal, in white shirts and dark suits; their mother a smiling matron who said almost nothing; another woman whose age seemed somewhere between the sons and the mother; and, across the table from me, a girl my age, as beautiful in the same way as her brothers were handsome. I asked her to dance before Hushpuppy did.

We walked side by side onto the dance floor. Her beauty made me too shy to take her hand. Besides, those two brothers were watching. As if on signal, the band played melodies fit for dancing slowly. We turned to each other to dance in the formal way, our left hands clasping, her right hand on my shoulder, mine around her waist, our bodies close all along their lengths. I told her my name. She told me hers: Wanda Maria Carolina Colombe. A name as lovely as she was. We didn't have to say much to each other to understand it was permissible to be in love just for the Fourth of July weekend, and that our time together would be intense, so respectful of each other, and to our futures with another, as to be chaste and unforgettable. After it was over, we'd never see each other again.

We danced for a while and then rejoined her family. We had strawberry shortcake and very black coffee, and then we danced some more under the watchful eyes of the brothers. Hushpuppy danced with the woman who was slightly older. I don't know if he told her there was a Mrs. Hushpuppy. She told him she was divorced, free at last, ready for fun. When the patriarch stood up from the dining table, everybody did too. They invited Hushpuppy and me to join the family tomorrow for a pig roast on the beach.

We went back on the same boat that had taken us out to Water Island, driven by the same young Black man. The steel band came with us. Halfway there, the boat's engine stopped. No sounds now, except the sibilance of tiny waves against the gliding hull. The half-moon glittered on the water. The boat driver must have decided that something was caught in the propeller, because he took off his shirt and dove in.

I did too. Why not? The water was warm; the moon was out. But I didn't take off my shirt nor my suitcoat, not even my white buck shoes. They'd dry soon enough. My yellow tie floated perpendicular to me on the surface, as if trying to get away. Apparently, the steel band decided this was a good way to spend the time until either the boat was fixed or wasn't. They jumped in too, and so did Hushpuppy. The boat driver laughed and laughed. "It's fixed now," he said. We floated around for a while on our backs, gazing up at the stars, and at the top half of our feet out in front of us, like ornaments in the bow of a little boat. Then we climbed aboard, the happy boat driver started the engine, and we went the rest of the way.

The same boat driver drove us to a yellow beach the next day. Blue sky, the taste of rum and lime, succulent meat, the smell of smoke. I sat next to Wanda, still under the careful and yet friendly eyes of her brothers—friendly to the boat driver too, who sat among us on the sand. He sang songs in Spanish that Hushpuppy and I could only listen to. When it was time to leave, we waded out to where the boat was anchored, and the driver took Wanda's elbow from me and helped her into the boat.

That night, at the hotel on Water Island, after the dinner was over and the dancing had begun, two very drunk chief petty officers in their liberty dress blue uniforms showed up. They were not from the *Vermillion*. They were louder than the steel band's music, swearing, dropping the F bomb more than once. When the boat driver emerged from the kitchen carrying a tray, they addressed him as a nigger. The two brothers came across the dance floor to Wanda and me. I said, "Goodnight, Wanda, see you tomorrow," and they escorted her away to safety. Everybody in her family, including Hushpuppy's dance partner, departed the scene. So did the young boat driver.

He returned seconds later with a pistol. He pointed it at the two drunk petty officers. The music stopped. If I were totally sober, I would have been a coward, but I had been made brave by Rum Collins. ("The juice of two limes will keep you sober enough to know what's going on no matter how many you have," Grizmala had informed me, "as long as you don't let them put any sugar in.") I walked across the dance floor with my hand out, keeping my eyes on my friend the boat driver's face, never looking at the gun, while he was deciding whether he hated these two white assholes enough to spend the rest of his life in jail. He handed me the gun. I have no memory of what I did with it. Nor any knowledge of where the CPOs went when they hurried away.

The next day, Hushpuppy and I went to the place we had agreed to meet Wanda and his divorcee dance partner to watch the Fourth of July parade. They didn't show up. I'd never see Wanda again.

After the parade was over, Hushpuppy and I went back to the ship, got our bathing suits, and headed for a beach. On the way, passing a space which I dimly remember as a large empty blacktop, like an abandoned airfield, we crossed paths with Noel, from Riverside. He was a friend of my brother Henry, and his parents were friends of our parents. His sister and I had been classmates in Riverside Public School. He was in the Marines, flying small planes to guide the aim of artillery on the ground. We chatted awhile, amazed at the smallness of the world. Then he went to where he was going, and Hushpuppy and I continued to the beach.

Several months later, I went home on leave. I took the train from Norfolk to New York, then a commuter train to Stamford. Noel's sister and her parents were on the platform, as if they had come to greet me. I told them how glad I was to see them, and that Noel and I had run across each other in St. Thomas. They listened and then they told me they were waiting for his coffin to be unloaded from the train. He'd crashed into a mountain several days before.

CHAPTER SEVEN

It was during that same leave that I persuaded Joanna that we should marry while I was still in the Navy, instead of waiting until my hitch was over and I had found a job. I didn't want to watch my married peers rush happily down the gangway into the arms of their wives at the end of long cruise even one more time. We planned that I would put in for ten days of leave, starting a day before Thanksgiving. If granted, we would get married on the Saturday after Thanksgiving.

My parents were horrified. Even in good times, their generation waited to marry until the husband had a secure job and a savings account. My parents were even more convinced that we should wait because they themselves had married in 1929, just days before the Wall Street Crash and people started jumping out of windows. My father was out of a job for several years. He was thirty-six when he married my mother. "You're still a child!" he blurted to me, and, immediately regretting, apologized. He must have known that what I was doing in the Navy wasn't children's work. Maybe he would have not apologized if he'd known how poorly I was performing.

But as soon as he realized he couldn't dissuade me, he and my mother became entirely supportive. They loved Joanna. I'm sure they continued worrying for her as much as for me, but they kept it to themselves. I never heard anther word of objection.

From then until we would marry, the *Vermillion* was at sea a lot in rough weather. When she steamed light—that is, without any cargo in the holds to serve as ballast—the combined weight of twenty-four landing craft on the main deck and in the davits above the bridge made us top heavy and thus very tender. Navy speak for "tippy." When we steamed broadside to the waves in rough weather, not just spray but solid water came over the rail and broke on the deck like surf on a beach. The ship would roll way over, almost on her beam ends, and stay there, draining water through her scuppers, then roll back, accelerating the other way. I remember one evening at dinner in the wardroom, I had lifted a coffee cup from its saucer just as my chair slid away from the table all the way to the bulkhead behind me, where it stayed for several seconds, my coffee cup still in my hand, slightly above me. Then my chair made the return journey depositing me back at the table, and I put the coffee cup, still full, not a drop having spilled, back in its saucer.

I was one of the lucky ones who never got seasick. Most of those who did got their sea legs after a day or so on the water. One summer night as we headed out under a full moon into mild swells, my boss, Lieutenant Lucky L., standing on the port wing of the bridge, as officer of the deck waxed poetic to me, his junior officer of the deck, about the moon's glitter on the water, the warm salt air, and how he loved going to sea. I listened with no comment. Why ruin his mood by confessing I hated going to sea, couldn't wait until this cruise was over, and had been counting the days since I'd reported aboard until August 5, 1955, when I'd be a civilian again? He stopped talking and approached the rail. I knew what was going to happen next. I'd seen it before. So, I sauntered away into the wheelhouse to give him privacy. Besides, I don't like to be around people who are throwing up. A few minutes later, his stomach empty, he was fine.

Poor Ensign Nelson B. was not so lucky. He was seasick every minute of every day and night we were at sea, no matter the weather. His tiny stateroom, shared with three others like mine, was on the other side of the ship from mine. In a narrow passageway just before entering it, there were drawers, as in a bureau, sunk into the bulkhead in the inboard side of the passageway. Once, in the rough seas off Cape Hatteras, I found Nelson crouched with his back against the outboard bulkhead, gathering the seasick pills and golf balls that rolled toward him after having fallen from one of the drawers that had slid open. He was grabbing as many as he could and stuffing them into his pockets before they rolled away from him again.

Nelson was the gunnery officer. His general quarters station was at the top of the house, the highest point on the ship from which to direct the fire of our several twenty-millimeter guns and our biggest gun, a five-inch-thirty-eight on the main deck, aft of the house. Somewhere off the coast of Virginia a small propeller plane visited us, towing a target to provide us gunnery practice. My job as the officer of the watch in CIC was to communicate with the pilot over the radio.

Nelson, dizzy and nauseous as always, told me to tell the pilot to make a run along the starboard side. I radioed the instructions to the pilot and several minutes later he flew along our starboard side from stern to bow, several thousand feet above us, the target streaming safely far behind. Nelson gave the order and the guns shot. Nothing hit the target.

"All right," said Nelson between dry heaves over the phone to me, "tell him to make a run along the port side." I gave the instructions as before. The pilot, maneuvering in preparation for the run, wanted to be sure all the bores were clear. A reasonable request, as sometimes a shell would stay in the bore of a gun to shoot out at an unpredictable time later—like when the plane, not the target streaming way behind, was in its trajectory.

"Nelson," I said into the phone. "Are all the bores clear?"

"What?"

"Are the bores clear?"

"Yeah, yeah, bores are clear."

"Affirmative. All clear," I radioed to the pilot. And he started his run.

BANG! An instant of stunned silence during which I was sure the pilot and his plane were no more. Then, to my enormous relief, I heard his F bomb crackling in my earphones. "Clear? My ass. That shell went by right in front of my propellor!" He dropped the target into the ocean and headed for Virginia. A few minutes later, I could barely hear him asking me for his course back to his base. He'd forgotten to ask, and I'd forgotten to tell. By that time, he was way beyond our radar. I had no idea where he was. Nor did he. "Just fly west until you see land," I suggested.

He didn't answer.

* * * * *

IN THE LATE SUMMER OF 1954, we were sent up Chesapeake Bay to shelter from a hurricane. The wind was so fierce that we dragged anchor for several miles despite using both anchors and steaming ahead one-third speed. When the storm subsided, we brought both anchors up to return to port. The starboard anchor, which was the one that was always dropped, had scooped a thick coil of wire off the sea bottom. I sent for a member of the repair division to cut through that wire with a blowtorch. Just as he started, I warned him to be careful not to burn a hole in the shackle that holds the chain to the anchor. "What did you say?" he asked, turning his head to me as he burned a hole in the shackle. For the next six months, every time the captain ordered me to drop the starboard anchor, I dropped the port anchor. He never noticed. If he had, he would have yelled at me again.

We continued to be at sea most of the time until November of 1954. Maybe that was why I had forgotten to put several close friends on the invitation list for Joanna's and my wedding. It never occurred to me to write them a letter and tell them. Most of them are dead now and so it is too late. And a few of the friends I didn't forget to invite were on ships at sea. One of them, Don Robertson, still a close friend, told me later that while I was getting married in Weymouth, Massachusetts, his destroyer, the USS *Hitchcock*, was riding out a storm in the North Atlantic. The waves that came over the rails onto the deck had flowed down the air vents.

While we were on an operation which would keep us at sea until only a few days before Thanksgiving, I realized I had forgotten to get a Wassermann Test, which was required in those days to prove you didn't have a venereal disease to be married. I got the test immediately after we returned to port, but how to get the results to the authorities in Massachusetts on time? It so happened that one of my closest friends since boyhood, John B., was a Navy pilot who by luck happened to be temporarily stationed on an aircraft carrier that was temporarily in port at the Norfolk Naval Base. How cool was that? He flew the results of the test to Boston in his jet. I got my permit just in time. When I asked John how he got permission for the flight, he said, "Don't ask."

I got leave for ten days, starting the day before Thanksgiving. Joanna and I were married on the Saturday after Thanksgiving. We had seven days for the honeymoon and one day to drive back to Norfolk.

CHAPTER EIGHT

Joanna and I woke up on the first morning of our honeymoon in a borrowed northern Vermont ski cabin to discover we were snowed in—the ultimate bliss for newlyweds. Except that we hadn't gone grocery shopping yet. There were several cans of stewed tomatoes in the kitchen cabinet, canned ham, some pancake batter, cooking oil, and maybe a few other edibles. Hardly a celebratory diet. We were at the end of a long driveway and our car was buried up to the windows in snow.

The family from whom we borrowed the cabin had hired a local resident to care for the place in their absence. He had lit a fire in the wood-burning furnace in the cellar, so the house was warm when we arrived. But now it was freezing. Down in the cellar, I put some more logs in the furnace and went back up to the living room to discover there was no running water. I called the caretaker, who confessed that he'd forgotten to top off the tank in the attic, but no worry, all I had to do was turn on the pump.

"Good, I can do that," I said.

"Yup. It ain't hard."

"Can you tell me where the pump is?"

"Yup. It's easy to find, ordinarily."

"Ordinarily?"

"Might be kinda hard to find in all this snow."

"Really?"

"Yup. Mighty early for snow."

"You can say that again."

"Don't need to. Once is enough."

Silence.

"The pump is about a hundred yards from the left corner of the front porch. And about twenty yards from a big tree."

"No shit!" I wanted to say. "It's fucking Vermont. There're a billion trees."

"You'll need to bring a shovel."

"You mean it's—"

"Yup. Snow shovel."

"Sunken?"

"Yup. But there's a cover."

"Oh good. That helps."

"Yup. Wouldn't want snow to get on the pump."

"So, I just turn it on? There's a switch?"

"Yup. Big red one. You can't miss it. Then you got to open the valves. Just turn 'em to the left."

"I know how to open a faucet! Why don't you just come over?"

"Take me a while. Probably by day after tomorrow. Roads'll be plowed by then. Soon as they are, I'll come by in my rig and plow you out. Cost you ten bucks."

"Well, I just hope I can find the damn pump."

"Me too. Shovel's in the cellar. And there's a couple buckets down there."

"Buckets?"

"For melting the snow on the stove if you can't find the pump." Then he hung up.

"There's no coffee!" Joanna said when I put down the phone.

That was the worst news of all. Maybe my dad was right: don't get married until you have a lot of money to spend on a Caribbean Island hotel instead of a borrowed cabin in the boondocks. At least it was a little warmer in the house now that the furnace was lit. I found the shovel in the cellar next to two buckets, which I prayed we wouldn't need. Then I went out onto the front porch and into a Christmas card scene: blue, cloudless after-storm sky, bright, spangling sun on virginal snow, the scent of the wood smoke wafting from the chimney. I felt a surge of energy and optimism. Joanna and I were *married!* I'd get the water turned on; we'd have ham and pancakes for breakfast. Stewed tomatoes for lunch. Ham again tonight, and maybe by tomorrow the roads would be plowed and we'd go shopping for groceries. Rent some snowshoes. Take the hikes we'd planned. Go out to dinner.

I stepped off the porch, waded through hip-deep snow for what I estimated was a hundred yards on a straight line from the left corner of the porch, and started digging. Several hours later, my shovel scraped on the wooden cover of the pump box. I was pouring sweat. The cover was frozen on; I pried it off, opened the valves, and turned the big red switch to ON. And heard the rumble of the pump. Success! I waded back to the cabin. By now it was noon. We had brunch. Ham and stewed tomatoes. We'd save the pancakes for breakfast tomorrow, though there was no syrup and no butter.

That evening we found a bottle of sherry in one of the cabinets—not our favorite beverage for sure, but given the circumstances, an elixir. We drank the bottle dry and ate the ham, and decided to listen to music on the old fashioned victrola. But where were the records? Joanna finally found one. That's right, just one. *So, let's not play it now. Let's save it for later in the evening.* Joanna put it on a chair near the victrola. Now, though I'd put several logs in the furnace, it was getting cold. *So, let's light a fire in the fireplace.* I did, a nice roaring blaze to accompany the sherry's buzz. The chair with the record on it was near the fireplace. Joanna, wanting to get warm, sat down—right on the record, and broke it. And the room filled with smoke because I forgot to open the flue.

We went to bed—always a wise decision, especially on your honeymoon.

The caretaker's prediction was inaccurate: the roads were plowed by midday the next day. He showed up in the afternoon and plowed us out—for fifteen dollars. "I forgot the driveway was so long," he said, explaining the extra five. Joanna and I decided we'd had enough of rustic cabins, even though we could stock it now with food and drink. We checked in at a nearby ski resort. It wasn't open for skiing yet, because no one had expected early snow, but after our foodless sojourn in the ski cabin, the resort's dining room and our bedroom, both warm and free of smoke, seemed luxurious.

For Joanna and me, there are only two memories of the place that persist: the taste of Cambell's mushroom soup, piping hot, prepared by someone else and fortified by sherry; and us sitting on the double bed in our room casually perusing the contents of my overstuffed wallet. I was almost as curious as she; there were compartments which I had not perused for years.

Joanna slid her forefinger in the wallet, exploring. Photographs maybe? I shrugged, not knowing. She extracted her forefinger, along with two small photos: one of my dad in a baseball hat, at Kidney Pond Camp near Mount Katahdin in Maine, grinning, holding a trout he'd caught in front of himself for the camera to see; and one of a beautiful girl, high school age, just her shoulders and face addressing the camera. It was signed, *Always, M.*

Joanna likes to kid me about finding a picture of a girl, signed *Always*, on our honeymoon. But she would no sooner ask me to tear my former girlfriend's face in two and relegate it to a garbage can than she would ask me to rip apart my father's grin. She slid both photos back into their compartment of the wallet. Both have disappeared long ago to I don't know where, victims of many moves. But the photo of my father and the trout had several duplicates, all of which were enlarged, one of which my mother gave me. That photo now resides on the library table in Joanna's and my house in Oakland. Above my bureau in another photograph, my father smiles in black and white at my mother, who smiles right back at him. They've been dead for years.

* * * * *

JOANNA AND I GOT BACK to Norfolk after dark on Sunday, December 5. In the morning the next day, the *Vermillion* would go to sea until Friday, December 17. I had rented an apartment not far from the naval base that had one bedroom, a tiny bathroom and a combined kitchen-living-dining room. Compared to my quarters on the ship, it was luxurious. It did not occur to me, however, that this would not be a happy place for my wife, who knew no one in Norfolk. For her, it was like being abandoned in a strange place for almost two weeks. I opened the door and turned on the light. She took one step in and started to cry.

In the morning she drove me to fleet landing where I'd take the liberty boat out to where the *Vermillion* was anchored, ready for departure. We noticed a snow flurry but were too distraught about being separated to think about the weather. Besides, we were from New England, where everybody knows how to drive in the snow. But now we were in Virginia. The *Vermillion*'s cruise was canceled out of concern that many of the crew would not be able to get to the fleet landing. Saved by a few snowflakes, I'd get home from the Vermilion to Joanna every night, except when on duty.

That Monday evening, joyfully on shore, I went shopping for groceries with Joanna at the naval base commissary. The lady at the checkout counter saw my name on my ID card. "Mr. Davenport! How wonderful to meet you," she said, practically jumping over the counter to hug me. "I'm Mrs. Grizmala. Thank you for being so nice to Richard!"

CHAPTER NINE

On Friday December 17, the same day the cruise would have ended, I received orders to report to the shore patrol office in Norfolk starting on Christmas Eve. I would be in charge of what in civilian parlance would be the night shift until dawn on New Year's Day. As I understood it, the procedure was to select a junior officer serving on a ship and assign him this temporary duty. So what if I didn't have the foggiest notion of Shore Patrol procedures, methods, and responsibilities? I'd be at the mercy of the chief petty officer in charge. The Navy needed to have an officer in charge to court martial for poor judgment or dereliction of duty when something went sour, which anyone past the fourth grade and not brain dead could predict as inevitable during the holiday season, especially on New Year's Eve.

Because of the season, most, if not all, the ships were in port. Thousands of sailors would go ashore every evening. One of the most disgusting places on earth would be fleet landing, late at night. Every other sailor waiting for the liberty boat back to his ship would be drunk and throwing up. Norfolk's menu of entertainment for sailors was limited mostly to drinking. If you were to give the world an enema, the saying went, Norfolk is where you'd put in the hose.

Even for officers who did their drinking at the Officer's Club, it was dreary, at least to me. I had spent many evenings there when the *Vermillion* was in port, before I was married. I can remember having only one interesting conversation there.

It went something like this. The person to my right at the bar said, "You're out of uniform." Technically, he was correct. I was wearing the khaki trousers of my work uniform, along with a civilian sport coat, shirt, tie, and loafers on my feet.

I ignored him.

"You didn't hear me?"

"I heard you," I said, looking straight ahead, not at him. I was sure this guy was a Marine, though I couldn't confirm it. There were plenty of assholes in the Navy too.

"Look at me," he said.

I shook my head. His hostility radiated. I felt my irritation expanding like a gas in my ribcage.

"What, you're telling me you're *not*?"

"I'm not telling you anything. I'm not talking to you. Can't you tell?"

"Just look at me and tell me those aren't uniform pants."

I turned. I'd expected pure, unadulterated ugliness. Something to feel sorry for him for. But he was good-looking. A movie star type. Blue eyes, high cheekbones, square jaw, cleft chin. I had a sudden powerful desire to reach for his nose, pinch it between my thumb and forefinger, and twist it off his face.

"See, that wasn't so hard, was it? All you have to do is apologize for being out of uniform and then go away."

"I'm not going away," I said. "But I'll buy you a beer if you shut your face."

"If I what?"

"You heard me."

He didn't answer. I wondered if he was about to swing a punch. I was ready to catch it. Like you catch a ball. And slam it on the bar. I was also on the verge of informing him that if he didn't shut up, I'd rip his head off and shove it up his ass so far it would come out his scalp. I actually formed the words in my head. It would be so satisfying to deliver such elegant repartee! Should I? Or should I just walk away?

While I was deciding, I heard someone slide into the chair to my left. I didn't turn my head to see him, but I heard him whisper in my ear, "He's got a black belt in judo." He paused to let this sink in. Then,

"Besides, he's crazy."

I was struck by the wisdom of this implied advice. Very sage indeed. Everybody knows the crazy guy always wins. So, I said goodnight to my adversary, thus saving a portion of dignity, and walked away from the bar.

Behind me, a punching sound. Like a wooden mallet on a grapefruit. Then the crash of a body hitting the floor. I turned around. A short person, dressed in a blue blazer, gray flannel pants, and shiny cordovan shoes, the one who had provided me such good advice, was standing over a tall, prone unconscious body. He must have had to jump to land the punch.

* * * * *

MY MEMORY OF THE SHORE patrol office in Norfolk is of brown walls covered with huge black-and-white photographs of car crashes featuring maimed bodies. A ghoulish place to spend a long night, but, as the chief in charge explained when he saw me gaping, they were there to motivate the Shore Patrol personnel to keep drunk sailors from getting behind the wheel by imprisoning them in the brig until they sobered up.

Which the Shore Patrol did in large numbers whether or not the sailors had access to a car. Most of them didn't, of course. It would be hard to buy a wheelbarrow with the pay these guys, many still in their teens, were receiving. It seemed as if the entire US Navy had convened in Norfolk for the purpose of becoming drunk and disorderly.

Early during my first night, a pair of shore patrolmen brought in a drunk sailor. They shoved him hard against the counter behind which another shore patrolman stood to register him and take his possessions— wallet, watch, cigarettes, and lighter—for safekeeping before putting him in a cell. With a stream of profanities, the drunk objected to being shoved, a reasonable protestation it seemed to me. One of the shore patrolmen responded by bending him over face down on the counter and holding him there while he struggled.

"Wait," I said, "you can't do this!"

The chief looked at me and shook his head. I was just one more ensign who needed to learn how things worked. Nobody else even appeared to have heard me. They took the drunk guy down a corridor and put him in a cell. It was around a corner, so I didn't see whether they threw him in or just gave the needed shove.

"You let them do that again, I'll report you and everyone in here, you understand?" I said to the chief. I was probably angrier at my authority having been ignored than at their mistreatment of the drunk. At any rate, before the full sentence was out of my mouth, I felt powerless. I was just some guy in an ensign's uniform who they had to ignore to do their job. Pandemonium was evolving in hundreds of bars and semisecret whorehouses in the dingy parts of Norfolk. Soon it would arrive at the door.

"Aye aye, sir," the chief said. Then he added, "As long as you say so."

I had the feeling I wasn't going to say so very long, and I was right. The next sailor they brought in was a giant. Vastly more disorderly than drunk, he struggled to get free of the two brawny shore patrol guys, one on each side of him.

"What now?" the chief asked me.

I didn't answer.

The two patrolmen dragged the man up to the counter, still struggling.

"Wait," the chief called. And looked at me. "Sir?"

"Just do it without bending him over," I said, finishing the sentence just as the man got his left hand free. He swung at the patrolman on his right side, missing. That patrolmen pinned his arms while the one behind the counter reached over and grabbed him by his head, using his ears as handles, and pulled while the other two shoved him against the counter and slammed his face down onto it. All three leaned hard down on his upper back while the chief stepped forward and emptied the man's pockets, placing the contents on the counter. I stood there, watching. It took all three of the shore patrolmen to get the still struggling gigantic sailor down the corridor and around the corner. I heard the cell door shut.

"Geez, how does a guy that tall get in the Navy?" the chief murmured to no one in particular. "He musta been lying down when they measured him."

Soon we were very busy—everyone but me, that is. I just watched. Often it wasn't just one at a time, but bunches. Around midnight, the chief said to me in a kind tone, like someone's uncle explaining, "You see, sir, it gets to be kinda like mass production, you know what I mean?"

"Yes, chief, I know what you mean."

"Well, thanks!" he said. "Some of the others take longer to—"

"Understand?"

He nodded. "Yeah. Understand. You want some coffee?"

As you have undoubtedly gathered by now, what I understood was that my job was to stay out of the way and do nothing, absolutely nothing, except sign the log at the end of the watch.

But on New Year's Eve, when the shore patrol headquarters in Norfolk was one of the busiest places on the planet, even I had to lend a hand. Just as 1954 became 1955, the chief, smiling all the while, asked me if I would listen to the charges against a sailor that an angry sex worker was claiming in a loud aggrieved voice. She sat up straight in a chair across my desk, trying to look like a citizen who deserved to be heard. It was obvious her hair wasn't naturally hers: it was blond, her eyes were brown. I was determined to listen respectfully. After all, I was an Oberlin graduate, proud that my college respected women by being the first to be coeducational, while Harvard, Yale, and Princeton were still essentially boys' clubs for WASPs.

"How can I help you?" I asked. I really did want to.

"You can make him pay."

"For—services?"

She didn't answer.

"I see. Do you know his name? What ship?"

"Bill Somebody."

"Bill Somebody? That's all?" I don't know why I asked. There was nothing I could do even if she could identify her customer. I should have just tried to soothe her. I'd only read about sex workers. Raskolnikov's in *Crime and Punishment* came to mind. This woman seemed more normal to me, much less exotic.

"They never give their names," she said.

"Then why?" I asked. A stupid question.

Her answer was that she'd had enough of getting screwed. She didn't recognize her own pun. And anyway, maybe that wasn't the service she had rendered to the cheater but some other that I didn't want to imagine. She talked at length, saying the same thing over and over in a variety of ways and free of swear words: she was tired not just of getting screwed, but of everything. The best thing I could do for her was sit there and listen respectfully. I figured she'd get quiet after a while and just go away.

That's not what happened, though. Instead, two shore patrolmen took her by her elbows and walked her to the front door of the building and pushed her out the door into the night.

"Why?" I wanted to know. "She would have left on her own."

"Because that's how we do it," the chief explained. While the woman was venting to me, he'd called the civilian police and told them a woman was disturbing the peace near the front of shore patrol headquarters. Then he waited exactly four minutes before giving the signal for her to be pushed out, just as the civilian police squad car arrived to pick her up and book her.

I wanted to tell the chief how wrong that was, how cruel, but my authority to pass judgment was merely official. Besides, I had to admit, his method was efficient. It was my last night on temporary duty as the officer in charge. At dawn on New Year's Day, I shook hands with the chief, said goodbye, and went home to Joanna. She was still asleep. I postponed breakfast and got in beside her. The next day I reported back aboard the *Vermillion*. Soon we'd go to sea again.

Looking back now, I realize I remember many fewer drunk and disorderly Black sailors than white sailors dragged in, shoved against the counter, and deposited in a cell. Did the Black sailors have as much reason to be wary of military police as they surely had of civilian police and therefore stayed sober? Or did they gather where the shore patrol was afraid to go? I'll never know.

CHAPTER TEN

In January of 1955, the *Vermillion* was ordered to the Brooklyn Navy Yard to be refitted for operating in the Arctic. If I remember correctly, a major element of the refit was a propeller that would not be damaged by ice. It would take about six weeks to be ready, and then the *Vermillion* would join Operation Deepfreeze, near Thule, Greenland, which was scheduled to last beyond August 5, 1955—when my required two years of active duty would end and I would become a civilian again. Joanna and I would be separated for about five months. We ended our lease in Norfolk and she traveled north to live with her sister in Greenwich, Connecticut, where her sister was an elementary teacher at the public school. On nights when I didn't have the duty, I'd commute there from Brooklyn.

My first watches as officer of the deck, as opposed to junior officer of the deck or CIC watch officer, were on the cruise north from Norfolk to Brooklyn. For all the other line officers on the ship (except Nelson B., who was so busy being sick he wasn't good at anything either), steaming independently was a walk in the park compared to keeping station in a formation. But it was plenty challenging for me.

If conning a ship were like driving a car—hands on the wheel, foot on the accelerator or brake pedal—I would have been good at it. But conning a ship is physically an abstraction. You don't touch the wheel; you give orders to the helmsman. You control the speed by telling a different person what to ring up on the engine order telegraph, which tells the officer on watch in the engine room what to tell someone else to do. The relation in space between the ship you are conning and other ships as projected on a radar screen or captured visually are reported to you in numbers: points of the compass, miles or yards of range. For me, this translated into a feeling that I wasn't in control. Nothing was in my body; everything was in my head, and my head wasn't built for the job. For people who stood watch with me, including sailors in the First Division who served as lookouts, this translated as stupidity.

When I took my first watch somewhere off Delaware in calm seas, every man in the First Division donned a life jacket and climbed single file up the ladder to the port wing of the bridge into the wheelhouse, where, one by one, they saluted me, wished me good luck, then filed out onto the starboard wing and down the ladder. This lasted a good ten minutes. Almost all of them managed to keep a straight face. I didn't join the hilarity by returning the salutes. There was a lot of ship traffic up and down the coast. I was focused on the several ships in sight and on radar. If the range decreased and the bearing stayed the same, that meant we were on a collision course. Besides, not all the men were making a good-natured joke at my expense. Some of them were expressing contempt. Some were doing both. What I remember vividly is how glad I was that Captain Lay was not on the bridge at the time—that, and my surprise that I wasn't hurt by the contempt nor pleased by the good-natured kidding. I didn't care anymore how the men felt about me. That's the first step toward being a leader.

In the intervals of time between standing watches as officer of the deck, I spent hours filling out work orders for the upcoming refit and repair of many parts of First Division equipment. I didn't put in for a new anchor shackle, because that would require explaining the hole in it. Nor did I put in a work order for maintenance of the massive diesel air compressor, which had been resident on the starboard main deck forward for several months. It was worth thousands of dollars. The First Division sailors were grateful to Grizmala for procuring it, since it was much easier on one's body to chip old paint and rust from metal with the aid of an air compressor and mechanical chipper than to bang on it with a handheld one. But where did it come from?

"You don't want to know, sir," was Grizmala's answer when I asked him how he managed this coup. I should have known better than to ask. This wasn't a *Navy* compressor—it was *civilian*! What vast amounts of stolen Navy gear (or coffee, an intra Navy currency) he had to buy it with from whatever civilian owner, it was best not to promulgate. Since it was not recorded anywhere on some form in triplicate, it didn't exist. Yet there it was, its shiny yellow paint reflecting the wintry sun, while also adding to our starboard list—and crying to be explained to the navy yard's officialdom.

I would be the first to try to explain its presence. Where's the work request, officialdom would want to know, and how'd you get it? The question would go right up the chain: first to my boss, Lieutenant Lucky L., then to the executive officer, and then to our skipper, who, on the top rung of the hierarchy, especially didn't want the questions. "Permission granted," he said when I asked his permission to drop it overboard. He didn't ask the reason why. My guys lifted it up off the deck just like a Mike Boat: it swung above us in the air as the crane moved it over the frothing water below and the *Vermillion* listed still more to starboard. There was a reverent moment. Our chins pointed skyward. Then the boatswain's mate who was manipulating the boom used the tripping wire to remove the hook from the ring to which the carrying cables were attached. The huge yellow machine disappeared down into the deep sea—and the *Vermillion* lurched back to a mere three-degree list.

* * * * *

I WAS THE OFFICER OF the deck again when we steamed into New York Harbor, one of the largest harbors in the world, but I didn't have any worries this time because the pilot we had picked up at Ambrose Light was guiding us in and Captain Lay was on the bridge. I was free to absorb the experience.

Forget Fitzgerald's "thrilling returning trains of my youth" to

St. Paul, Minnesota. This was *New York*. So different when you arrive by ship than when you burrow under the ground by Grand Central Station. There was the Statue of Liberty to port welcoming strangers; in front of us was the narrow southern end of skyscraper-bearing Manhattan between the Hudson and the East River. New York Harbor smelled more like the ocean than oceans ever did. The tide was high. When we glided under the Brooklyn Bridge, it seemed there was very little room above our smokestack.

Joanna got a job with a New York City publisher as a secretary, though she was as well read as any male right out of college and could have probably done her boss's job. We both tried to prepare ourselves for the months of separation that would begin when the *Vermillion* slid back into the water and headed for Thule.

The Marine sentries at the naval yard seemed to have a particular animosity for the *Vermillion*'s sailors. Manhattan, just a subway ride away, was a nirvana compared to Norfolk for a sailor on liberty, but time after time the sentries refused to permit our sailors to go out through the gate. Instead, they'd send them back to the ship because their shoes weren't polished to a sufficient shine, or the part of the white T-shirt visible under the sailor's blouse contained Irish pennants—which meant, in Navy speak, that it was frayed. Some sailors were sent back to the ship twice and still weren't allowed to go ashore even after all the dress adjustments. I'm not sure the Marines had the authority to make those judgments, but they continued to harass our guys. Looking back now as I write this, I'm wondering why our skipper didn't use his authority to make the harassment stop. Maybe no one told him about it. I could have, but it never occurred to me.

One night, returning in civilian clothes to the *Vermillion* from a visit with Joanna and her sister, I showed my Navy ID card to a Marine sentry at an entrance to the navy yard. The sentry saluted and I walked through the gate. Less than a minute later, I heard loud voices and then a scuffling behind me. I turned around and watched a *Vermillion* crew member wrest the rifle out of the sentry's hands and club him with the butt in the forehead. The sentry slumped to the ground, and I turned around and continued walking.

The next responsible person to return to the yard would get medical help for the sentry, and I wasn't about to report the sailor, thus sending him to prison for who knew how long. Justice had been served.

CHAPTER ELEVEN

Early in February, a radioman handed me orders to report for duty at the Bureau of Naval Personnel in Arlington, Virginia. Shore duty! I laughed and threw the message away. I thought it was a joke, one that we loved to play often on each other. I'd received orders to almost every continent on the planet. Just the day before, I'd persuaded one of the radiomen to type up some orders for Hushpuppy to report for duty at an uninhabited atoll in the South Pacific. Everybody on the ship knew I was a newlywed, so the news for shore duty was much too good to be believed.

But a few days later, the radioman who'd handed me the message was surprised to see me still on board. He insisted the orders were real. "You are supposed to report for your new duty at Bu Pers in a week." I was stunned by this good fortune. There'd be no months of separation from my wife. Instead of being at sea in iceberg-haunted waters for almost six months, I'd spend eight hours a day in an office, have weekends off, and be home every night with Joanna!

Joanna quit the job she'd had for only a couple of weeks and I said goodbye to my envious shipmates. We packed up my secondhand 1951 Plymouth with everything we owned—including Joanna's goldfish in its aquarium, stored on the floor in front of her—and headed for the Washington, DC area. Somewhere along the way, Joanna accidentally kicked the aquarium, spilling the water and the goldfish out onto the floorboards where the poor little thing flapped desperately. But our good luck charm had not expired: up ahead, a Howard Johnson's restaurant beckoned.

I sped up, turned into the driveway, and screeched to a stop, and

Joanna scooped up the fish into the waterless aquarium and handed it to me. "Hurry!" she said, a totally unnecessary suggestion. The fish lay on the bottom of the aquarium, its gills opening and closing, while I dashed into the restaurant and up to the counter yelling, "WATER! WATER!" I handed the aquarium to a young man behind the counter. He calmly filled it and the fish, to loud applause, revived and swam again.

We found a furnished apartment in Arlington in a recently built building, luxurious compared to our Norfolk rental. We were in heaven, and so was the goldfish, it appeared, ensconced as the centerpiece of the dinette table. Joanna found an interesting research job with a government agency—a less sexist industry than publishing—near Dupont Circle in Washington.

My job at the Bureau of Personnel was to render the service records of officers up for promotion into forms that abbreviated that content to just the essential elements, saving time for the selection committees. There were about thirty of us doing this. Our commanding officer was a lieutenant commander who wrote sea stories for the *Saturday Evening Post* when he was off duty. Every one of us had recently been ordered ashore into this job, which was immeasurably easier than being at sea. We were all Reservists whose hitches would soon end. We were young, and the economy was expanding. The civilian world we were about to enter was our oyster. It didn't occur to us that one hundred percent of the oyster had been awarded to us because of the color of our skin. We partied almost every night. So what if we were tired and hungover in the morning? All we were going to do was sit at a desk.

One of the places we gathered to party was a swimming pool. As the weather started to warm in May, some of us brought bathing suits while others swam in their underwear. There was a lot of showing off from the high dive. We'd throw coins into the deep end and then would dive and compete to collect them. At precisely eleven pm, a young woman would appear in a window high up in the building next door and perform a striptease for us. It became so routine that we'd all get our drinks and gather in chairs a few minutes before. We'd all, dates and wives included, applaud and cheer as the woman revealed more and more of herself. She'd get almost entirely nude, having as much fun as we were, I'm sure, and then she'd turn the light out and disappear.

But wait! Just now, as I finished writing about our celebrated anonymous stripper, I Googled the deployment history of USS *Vermillion*, AKA 107, to get the exact dates of the ship's participation in the operation near Thule. I couldn't find them. Instead, I found a paragraph reporting that the *Vermillion* spent most of 1955 doing the same old thing as it had when I had been aboard: operating off the East Coast and in the Caribbean. Was this incorrect? Or were her orders changed after I left her? Did the Navy send her to drydock for months and who knows how much money so she could operate in the Arctic, and then send her south to the Caribbean instead? If so, she would have returned to home port in Norfolk after being at sea for operations lasting one month at most. That would've made Joanna's and my newlywed happiness at the gift of shore duty a bit less ecstatic, had we known that we'd been saved from being apart continuously for just one month instead of five months. But we didn't know at the time. Not an evening went by that I didn't think of my former shipmates, many of whom were married too, stuck on a ship, miles away on a frozen ocean, and rejoice that I wasn't.

* * * * *

AN ADDITIONAL DUTY OF OURS at the Bureau of Personnel was occasionally to deliver an important, confidential document to an officer of high rank—because, obviously, a mere enlisted person could not be trusted. Like maybe he'd sell it to Russia? One morning I was assigned to carry an envelope in a case to Admiral Rickover's office and hand it to him. He, of nuclear submarine fame, was standing in the middle of his office when I put it in his hand. He thanked me courteously and asked me questions about myself: where I'd gone to college, where I'd served before this duty. He was genuinely interested; it was clear to me that he did this with every young officer. Finally, he asked if I planned to stay in the Navy. "No, sir," I said, resisting the temptation to add that I had no desire and that I didn't think I'd be good at the work. If I had and he pressed me for more, I would have told him about the near collision I'd caused. But he didn't. The conversation lasted less than five minutes, sixty-nine years ago as I write this. I still don't know what it was, beyond his genuine, unstudied interest in me, that made him unforgettable. Some natural authority, granted from above? Whatever that was, I do know that if he had run for president, I would have voted for him.

* * * * *

NATURALLY, A CERTAIN PERCENTAGE OF the service records we briefed were those of officers who were or had been Naval Aviators, sometimes referred to as "flyboys." From watching them through binoculars taking off from and landing on aircraft carriers and practicing dive bombing and pulling out, it seemed, just yards above the water, I had concluded that they weren't just fearless—they were crazy. Pilots don't *land* on aircraft carriers. They perform controlled crashes. I knew for a fact that my good friend John B., who'd flown my Wassermann Test results to Boston, was incapable of fear. He had successfully persuaded me to do dumb, scary things with him when were kids—like diving off the crosstree high on the mast of a yacht in the middle of the night into water that may or may not be deep enough, only because I was even more terrified of being nominated a *scaredy-cat*. One time when we were riding a roller coaster at Playland in Rye, New York, he stood up at the top of the highest rise and pretended to be a tourist guide, delivering absurd information about items in the landscape, as we made the long, stomach-overturning dive to the bottom. I remember I squinched down in the seat and held on to the safety bar with all my strength.

In our family album there is a black-and-white photo of his jet flying above the Hudson River. On the reverse side of the photo was John's note: *I was planning to fly under the George Washington Bridge, but I changed my mind. I realized someone might be watching.* Time and time again in flyboys' service records, I came upon recorded official notations of reprimands for doing something really dumb that, for the miscreant, was probably a whole lot of fun. The one that stands out is a nighttime hop in which the pilot turned one light on and flew very low, straight toward an oncoming freight train, so he looked like the engine of a train on the same track heading for a collision. Of course he pulled upward just in time, zooming just a few feet above the terrified freight train driver.

One night when the *Vermillion* was in port and I had missed the liberty boat to fleet landing, Captain Lay courteously invited me to ride with him in the captain's gig. I engaged him in conversation so that he would not glance back over his shoulder at the *Vermillion* and realize she was tethered not to the starboard anchor but the port. We passed by an aircraft carrier, and it made him think about his classmates at the Naval Academy who had become Naval Aviators. He told me that not one of them was still alive.

* * * * *

A WEEK OR SO BEFORE my two years of active duty were to end, I learned that Captain Kirk, the *Vermillion's* beloved skipper, was among a group of captains eligible for consideration of promotion to the rank of rear admiral. I asked to be the person who summarized his record for the selection committee. I found this:

Navy Cross
AWARDED FOR ACTIONS
DURING World War II
Service: Navy

Division: S-42 (SS-153)

GENERAL ORDERS:

CITATION: The President of the United States of America takes pleasure in presenting the Navy Cross to Lieutenant Commander Oliver Grafton Kirk (NSN: 0-62655), United States Navy, for extraordinary heroism in the line of his profession as Commanding Officer of the U.S.S. S-42 (SS-153), in action against an enemy Japanese cruiser on 11 May 1942, during the FIRST War Patrol of that submarine in enemy controlled waters of the Southwest Pacific. Although Lieutenant Commander Kirk was operating under adverse and dangerous conditions due to mechanical deficiencies existing in his ship, and in spite of great physical strain from a long submerged patrol through enemy infested tropical waters, he adeptly maneuvered his ship to deliver three torpedo hits into the enemy cruiser which countered with a severe depth charge attack. The S-42 succeeded in escaping safely from this engagement after setting the cruiser afire and sinking her. This action, in which Lieutenant Commander Kirk displayed outstanding courage, endurance and resourcefulness, was in keeping with the highest traditions of the United States Naval Service.

* * * * *

THE LAST EVENT BEFORE BEING discharged from active duty in the Navy is getting a physical exam. The doctor who examined me was a Reservist called back to active duty because of a shortage of doctors. He was a captain, an unusually high rank for such duty. Like every normal male human, I dreaded the part where I would bend over and the doctor would stick his vaselined forefinger up my rear end to palpate my prostate gland where cancer often starts. But my recent good luck continued right up to this last event in my Navy career. "I'm a four striper, damn it," the doctor said. "And nobody is going to make me stick my finger up anybody's ass, including yours." I told him that was fine with me.

A few minutes later, I was a civilian.

POSTSCRIPT

About a year ago, my grandson asked me what life was like aboard the *Vermillion*. I typed USS *Vermillion* into my phone and up came several pictures of the ship—and also a video of her about to be sunk to make an artificial reef off the coast of South Carolina. I was surprised at how disturbed I was, especially because a reef is a good thing to become for a ship who's outlived her time. She'd been stripped of all her superstructure. All that was left of her was her hull, floating unanchored, waiting to die. Huge square holes had been cut in her sides, exactly where we'd clambered down the embarkation nets and slammed the radios. We had tried so hard to keep her safe, and now she was being sunk *on purpose*.

My grandson and I watched huge billowing flames and smoke burst out of the *Vermillion*. Explosives had been planted in her to speed her death. She started to slide under the surface, stern first, her bow pointing toward the sky, and slowly disappeared, leaving fountains of water and spray on the surface as air escaped upward out of her. And then she was gone.

Another video came on, praising the virtues of vacationing in South Carolina where the scuba diving is quite special. Fish, striped bass among them, explore the passageways we used to travel on our way to general quarters stations, while in the background a singer sings about jumping right in.

And then no other than John Denver sings about *Carolina On My Mind*. I was especially entranced by the underwater closeup of the huge winch for the anchor chain, while he, who is also no longer with us, sang on and on.

ABOUT THE AUTHOR

My first full-time job, after graduating from Oberlin College in 1953 and then spending the two year in the Navy as a ninety-day wonder, was as a Wall Street banker. I chose that work because my wife and I wanted to live in NYC and go to theater as often as we could afford, and also because it was an acceptable screen for a young, privileged WASP to hide behind while I wrote the Great American Novel.

By staying up late every night, I managed to produce the most self-indulgent, sophomoric novel ever written—except by people were actually still sophomores—and I mean in high school! Somehow, it managed to get the attention of an editor at Doubleday, who wrote me a few pages on how I might rewrite it to make it publishable. But I soon found I could not rewrite it. In the time it took to compose the novel, I had outgrown its premise entirely. So I buried the manuscript under the shirts in my bureau drawer and decided to take some time off from writing to catch up on sleep.

But I couldn't sleep. Because now without the novel to think about, I thought about how ardently I didn't want to get up in the morning and spend the whole day pretending to be a banker. I was amazed that no one caught me out. I would have confessed: this is not me, but that didn't happen, and I was soon promoted up the next rung on the young executives' ladder. My parents were delighted and relieved. Their son was climbing upward along an acceptable path. But my college friends, when I told them I was a banker, either laughed or looked concerned. Soon I'd own a house in the suburbs, take the train to the City every morning, reading *The Wall Street Journal*, like a character in a story by John Cheever.

I started perusing employment advertising in *The New York Times* and happened upon one for a teacher of English and coach of football, basketball and track at a boarding school. The position had come open too close to the beginning of the academic year for the school to have time to interview more than a few candidates. That's why I got the job. It was one of the luckiest days of my life.

For the next three decades I served in independent day and boarding schools. They all had their own beloved, idiosyncratic cultures like that of the fictional Miss Oliver's School for Girls. I found that purveying my passion for literature to the still supple minds and hearts of teenagers was all the satisfaction I needed. I never had to ask: Why am I doing this? And besides, I never had to wear a suit!

I didn't have to do any research for the novels in the Saga of Miss Oliver's School for Girls: *Saving Miss Oliver's*, *No Ivory Tower*, and *The Encampment*. Nor for *Ninety-Day Wonder*.

I simply lived the life.

ACKNOWLEDGMENTS

Thank you to Linda Lancione, Julie Chagi, Kira Petersons and Ed McManis. We've been together as friends and editors of each others' work for decades. Their suggestions made *Ninety-Day Wonder* a better read by far.

Thanks also to Olivia Ngai for her excellent copy editing, and to Rachel Metzger for bringing The USS *Vermilion* alive on the cover and her fine interior design.

Other Books by Stephen Davenport:

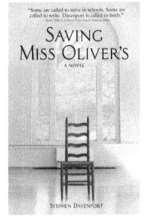

Saving Miss Oliver's
Book 1 of 3

No Ivory Tower
Book 2 of 3

The Encampment
Book 3 of 3
One of *Kirkus Reviews'*
Best Books of 2020

Printed in the USA
CPSIA information can be obtained
at www.ICGtesting.com
LVHW021355021123
762345LV00009B/72